Disasters and Life in Anticipation of Slow Calamity

The book provides insights into community narratives concerning life in the face of creeping calamities through a case study from the Colombian Andes. It sets out to make sense of the lived experience of disasters that are slowly unfolding as well as disasters that have not yet occurred.

This book explores what it means to live in anticipation of disaster and in anticipation of an uprooting of community, sense of self, and sense of belonging. It questions whether community resilience is a useful concept in the context of slow-onset geological hazards for which few viable solutions are available. The book forces us to think about how resettlement and displacement functions in the context of slow calamities, which presents distinct challenges, mainly related to lower political saliency than is usually the case in emergencies. The book thus also has implications for how we think about the adverse impacts of climate change. By raising new questions on the nature of disasters and calamities and how we experience them, the book explores the challenges and tensions surrounding governance and governmentality.

The interdisciplinary blend of practice-oriented and conceptual reflections will appeal to academics in postgraduate and postdoctoral research in social sciences, specifically, disaster research, geography, and research fields centred on natural hazards and disasters.

Reidar Staupe-Delgado is an Associate Professor at UiT The Arctic University of Norway and a Marie Skłodowska-Curie Actions Individual Fellow (MSCA-IF) at Roskilde University, Denmark. His research interests revolve around disasters, health emergencies and broader social problems, with a particular focus on slowly manifesting disasters.

Routledge Studies in Hazards, Disaster Risk and Climate Change

Series Editor: Ilan Kelman, *Reader in Risk, Resilience and Global Health at the Institute for Risk and Disaster Reduction (IRDR) and the Institute for Global Health (IGH), University College London (UCL)*

This series provides a forum for original and vibrant research. It offers contributions from each of these communities as well as innovative titles that examine the links between hazards, disasters and climate change, to bring these schools of thought closer together. This series promotes interdisciplinary scholarly work that is empirically and theoretically informed, with titles reflecting the wealth of research being undertaken in these diverse and exciting fields.

For more information about this series, please visit: https://www.routledge.com/Routledge-Studies-in-Hazards-Disaster-Risk-and-Climate-Change/book-series/HDC

Disasters and Life in Anticipation of Slow Calamity

Perspectives from the Colombian Andes

Reidar Staupe-Delgado

Routledge
Taylor & Francis Group

LONDON AND NEW YORK

First published 2022
by Routledge
2 Park Square, Milton Park, Abingdon, Oxon OX14 4RN

and by Routledge
605 Third Avenue, New York, NY 10158

Routledge is an imprint of the Taylor & Francis Group, an informa business

© 2022 Reidar Staupe-Delgado

British Library Cataloguing-in-Publication Data
A catalogue record for this book is available from the British Library

Library of Congress Cataloging-in-Publication Data
A catalog record has been requested for this book

ISBN: 978-0-367-25508-4 (hbk)
ISBN: 978-1-032-10598-7 (pbk)
ISBN: 978-0-429-28813-5 (ebk)

DOI: 10.4324/9780429288135

Typeset in Bembo
by SPi Technologies India Pvt Ltd (Straive)

Contents

PART III
Reflections **87**

Figures

Foreword

On life, time and disaster

Reidar Staupe-Delgado's book is about life. This single observation makes it a worthwhile, in fact a critical exploration within disaster scholarship. Disasters indeed mirror normative expectations about life. Disasters draw a line between what is an acceptable life and what is not, the *dis-* in the *aster* (from the Latin *astrum*). All interpretations of the concept focus on this threshold between what a normal life is and the *dis-*, for example, in Fritz's words, 'a disaster is defined as a basic disruption of the social context within which individuals and groups function, or a radical departure from the pattern of normal expectations'.[1]

Expectations of what a 'normal' life is, and hence of when the threshold of the unacceptable is crossed, thus turning life into a disaster, are inherently subjective and mirror a tension that Canguilhem long ago flagged in the context of medicine. Yet, as Reidar Staupe-Delgado demonstrates in this book, these expectations have been normalised by scholars of disasters and, even further, by policy makers and practitioners of disaster risk reduction who emphasise quantitative thresholds or normative expectations about what constitutes both a 'normal life' and, alternatively, a disaster. This normalisation of the term responds to the injunction of Western discourses about the meaning of the term 'disaster'. Such injunctions continue to emphasise large and rare 'events', to the detriment of slow and creeping processes, and that are shaped by the historical prominence of physical sciences in disaster studies. One is only to look at the EM-DAT database, which has become the standard in the field, to find evidence of this normalisation of disasters.

The example of Aponte that Reidar Staupe-Delgado uses in this book is a case in point. It shows that the slow calamity that local people are dealing with fall between the normative expectations of disaster studies and disaster risk reduction as a field of policy and action. In fact, the study shows that, from the perspective of normative expectations about what 'normal' life is, Aponte seems like a thriving 'community'. Yet the slow and creeping disaster and the anticipated cumulated impact it may have is central to the life of local Inga people. This tension between, on the one hand, normative and allegedly objective interpretations of disaster by outsiders, and, on the other hand, the subjective experience of locals is essential to understanding both the limits of disaster studies and pitfalls of disaster risk reduction. This is a major contribution of this book.

Moreover, life is inherently about time and Reidar Staupe-Delgado's book is providing an account of life in the *longue durée*. As such, this book clearly distinguishes between the fragmented time of disaster studies and disaster risk reduction and the duration of life, in Bergson's term, or what Reidar Staupe-Delgado calls the '*lifespan of a slow calamity*'. Unlike time and its mathematical fragmentation into minutes and hours, months and years, the duration of life can't always be divided as in Western thoughts. Still, disaster studies and disaster risk reduction, including disaster recovery, is commonly paced by the nature of the physical processes, then divided into the stages of the celebrated disaster management cycle, and ultimately structured after the normative injunctions of (Western) planning and project management assessed after criteria and deliverables set at given points in time. On the other hand, life, as this book brilliantly shows, is about what's happening between these set points in time, between the start and end of set stages in the disaster management cycle. It is about the continuing experience of the everyday, of the every moment.

As such, the case of Aponte and the local Inga people's experience of disaster challenges our very (Western) assumption about time. It blurs the distinction between the past, the present and the future, in a Bergsonian perspective. It shows that people live in the future, that is the anticipation, but also in the past, that is the melancholia, to make sense of the present in a way that defies the conventions of Western disaster risk reduction. Conventions of Western disaster risk reduction inherited from the Enlightenment and the injunction to fragment and categorise to better control Nature, its hazards, and life. A process that obviously does not match the worldview of the Inga.

Reidar Staupe-Delgado's book is therefore an invitation to reconsider our interpretation of disaster away from the certainties of disaster studies, as shaped by Western science, and the normative injunctions of disaster risk reduction, as a field of policy and practice structured after Western expectations. It is a pioneering piece of research that challenges our assumptions as much as it reveals the need for reinterpreting what we mean by disaster, whether slow and long or quick and short. Reidar Staupe-Delgado's book is a call for looking at disaster from the perspective of life and duration rather than through the (Western) lens of events, losses and sequenced response. One may, in fact, foresee in this pioneering effort the precursory steps towards a brand new approach to studying disaster; an approach that is vitalist, in Bergson and Canguilhem's sense, and that may open new grounds for understanding disaster from culturally grounded perspectives.

JC Gaillard
The University of Auckland

Note

1 Fritz (1961: 655).

Preface

What is a disaster? This question has perhaps become a defining feature of the emerging field of disaster scholarship. Critics have pointed out that we continuously fail to reach a 'consensus' on the matter. Others have suggested that the answer to the question will vary, depending on our perspective and the organisation we represent. Still other reactions centre on the organic evolution of answers to this question. Indeed, it has been claimed that, in retrospect, advancement in our understanding of key concepts is in and of itself the best measure for scholarly advancement. Perhaps the question is: what one could refer to as essentially contested? In either case, my position is that consensus on the question is not a goal we should pursue but rather that continuous reflection on this question and the wider nature of disaster as a phenomenon is what will ultimately drive us towards better and more pluralistic understandings.

Voices emerging from (or sympathising with) critical realist philosophical approaches to research have stressed that our view of what constitutes reality is distinct from how we obtain knowledge about realities and make sense of them as part of lived experience. Pluralistic views on disaster as a phenomenon thus allows us to conceptualise disaster experience as something quite distinct from natural hazards. In accepting the view that hazards need not turn into disasters, we also implicitly render comprehensible the idea that how disasters are lived is only connected to the actual mechanics of the hazard, to the extent that hazards augment how the disaster manifests, whether it be its intensity, geographical scope, temporality or other hazard-related characteristics that have a bearing on how populations in harm's way experience the adversity that all too often follows.

Studying disasters from different perspectives and in different contexts expands the available body of scholarship precisely by continuously providing us with an increasingly nuanced understanding due to this pluralism. I would argue, therefore, that the field can be described as non-cumulative, but we may still consider each contribution as providing one part of the whole. However, as opposed to a jigsaw puzzle that may one day be completed once the final piece is found, disaster scholarship may find itself unable to stretch for a theory of everything. Instead, armed with a diverse set of perspectives on disaster as phenomenon, we may still find ourselves better positioned to provide relevant and critical analyses.

This particular book employs a narrative view of calamity with a particular focus on temporal aspects of the disaster phenomenon. The book's title, *Disasters and Life in Anticipation of Slow Calamity*, tells us at least two things about what it sets out to achieve. I decided to focus extensively on the notion of living in anticipation as a result of my observation that disaster experience in the present is shaped not only by present-day adversities, but also by past experience (e.g. in the form of traumas, loss) or anticipated future hardships. Even under conditions where present-day disaster impacts are of moderate severity, the alarming prospect of worsening future conditions still affects live in the present, although they have yet to become manifest.

The object of anticipation in the case study engaged in this book centres on a slow calamity in the Colombian Andes. I have used the term slow calamity here as a result of a controversy in the research field. Although the book will grapple with this controversy to a great extent, I want to quickly outline its essence. Disasters are generally conceptualised as a result of underlying socio-economic conditions. We have consistently observed that societal factors shape disaster outcomes to a much greater extent than the nature of the natural hazard phenomena who often get the blame in popular discourse. We thus rarely speak of rapidly occurring disasters, as the disaster risk creation process which ultimately produces disasters is inherently long, often centuries in the making. At the same time, some disasters are characterised by a near-instantaneous ruination of the built environment, whilst others impact communities much more slowly. Opting for the phrase slow calamity was thus a way for me to engage with this topic without unnecessarily colliding with the way disaster causation is understood in the field. Focusing mainly on what I refer to as the 'onset phase' of disaster, then, in writing this book I was mainly interested in accounting for how the onset dynamic of disaster impacts shape lived experience, with a specific focus on how present experience is overshadowed by the knowledge that even harder times await in the future.

Work on the concept for this book began in 2016, following a field expedition to Aponte, a town in the Colombian Andes. At the time I was working on my doctoral research, doing fieldwork in south-western Colombia on how El Niño is prepared for. This project was also essentially about the onset phase. My aim was, essentially, to study what happens from the moment at which an El Niño warm event is confirmed, up until the time at which it peaks. The doctoral project itself had as its main objective to better conceptualise the disaster preparedness concept, where my focus became particularly centred on the role of uncertainty and anticipation as central to understanding preparation. As I happened upon the slow calamity which serves as the topic of this book, these ideas naturally merged to form the analytical basis for this volume. It must be noted, however, that this book is not primarily centred on telling the story of Aponte. Its primary objective is to provide a reflection on the nature of slow calamities and the way in which they give rise to a sense of 'living in anticipation'.

At the outset I must admit that the process has had its challenges and the book is not free from limitations. The book's primary data are mainly based on an expedition from 2016, but supplemented with several follow-up visits to Pasto and Bogotá, where I had the opportunity to collect secondary data as well as participate in stakeholder meetings. Loss of contacts and changes in the security situation made longer fieldwork with the purpose of delving deeper into the topic challenging after my doctoral project had finished.

I was also surprised to find that the slow calamity affecting Aponte lasted much longer than I had initially expected. As this book is being written, the future of Aponte still remains unclear. The tendency for disasters to not manifest as expected or follow predictable patterns is for this reason also a key theme in this book. We may also fruitfully distinguish between not only the hazard phenomenon and the disaster phenomenon, but also between the disaster itself and societal responses to it (i.e. disaster management). Not only did the hazard phenomenon manifest more slowly than one might initially have expected, but the response to it has also become protracted. With resettlement and recovery existing only as elusive necessities that are yet unfulfilled, only time will tell how the situation will eventually play out. In late 2020, the resettlement process described in this book was gaining momentum, but the situation remained unresolved in practice.

At the time of writing, the on-going COVID-19 pandemic has claimed over 60,000 lives in Colombia. Central indigenous stakeholders and other political figures have succumbed to the disease or suffered illness. While the economy and labour market has not been hit as hard as first projected, leaders are making it clear that hard prioritisation will shape the plan for economic recovery. We may expect that the crippling effect of COVID-19 on Colombian society, economy and political agendas may further lower the attention devoted to the plight of the inhabitants of Aponte. In other words, although this book certainly sheds some light on the story of this community, speculating how things will end is beyond its scope. Instead, I offer at least two potentially fruitful contributions to the field by drawing on the story of Aponte. One is directly connected to sparking debates on the question posed at the beginning of this preface. The other is a potentially interesting analytical lens through which we can make sense of how looming adversities shape life in the present.

Acknowledgements

Every research endeavour invariably draws on the support, skills and input of many different people. As my first published monograph, this book is no exception. Whilst I have enjoyed writing this book far more than I expected at the outset, it has also been a far longer and more difficult process than I had first imagined. It turns out the hardest part about writing a book is combating procrastination and prioritising a long-term project despite numerous opportunities to take on more short-term projects. Although the topic of this book came to my attention by chance, there are many people who deserve thanks and recognition for having led me to this point, knowingly and unknowingly. This book is the culmination of many other related and unrelated projects.

My academic career started in 2014 when I was accepted onto the PhD programme in societal safety at the University of Stavanger in Norway. My academic career would perhaps not have happened, and neither would this book, had it not been for the potential that my doctoral adviser Bjørn Ivar Kruke saw in me. Thank you for starting me on this path and for encouraging me to follow up on the Aponte case despite my PhD fieldwork being centred on an unrelated topic. I also wish to thank other colleagues at the University of Stavanger for having supported me and discussed my ideas about the case with me during my PhD fellowship period. A special thanks to my dear friend Lisbet for having read and commented on an early draft of the first chapters of this book. The book is also a product of the unconditional support from Eva, Francesca, Lise and Sandre. I know you had almost given up on my book project, but thank you for always encouraging me to dedicate more time to finishing it.

I also wish to extend my deepest gratitude to the United Nations Development Programme (UNDP) office in Pasto for inviting me to participate in the field expedition to Aponte in 2016, and for allowing me to observe subsequent meetings. I also wish to thank the Inga for welcoming us and for underlining that they want their story to be read. At UNDP, I was kindly offered a working space, which gave me the opportunity to discuss my ideas with people familiar with the situation. However, I must underline that the views and interpretations presented in this book does not reflect the opinions of UNDP or any of their staff. Still, their support and openness was what led to my knowledge of the disaster in the first place.

At my PhD defence in 2018, I was invited to give a talk that problematises the concept of community resilience against disasters. I decided to use the material I had available after the field expedition to Aponte to discuss the concept of resilience. I particularly emphasised how resilience can be understood in contexts affected by perpetually worsening slow-onset hazards, using Aponte as an example. I also reflected on the relevance (or lack thereof) of the resilience concept in contexts where resettlement may be necessary, discussing whether resettlement may be considered a resilient move. My opponents, Prof. David Alexander and Dr. Kerstin Eriksson, provided interesting reactions to my talk, which helped me structure my thinking around the topic. The talk ultimately led to a publication on the same topic in the *Journal of Development Studies*. This book represents an effort to go more in depth, considering many more sides than the controversial concept of resilience, which, in turn, means that I had to delve much deeper into the subject matter to deliver on my promise to tell their story.

Many people have also contributed in direct ways. I am grateful for the support of the Colombian Geological Survey, who consistently provided me with documents, images and data upon request. I would similarly like to thank Felipe Molina for his help with technical translation, information condensation and his excellent consultancy work for parts of this book's data material. Further, I would like to thank Dominic Ochotorena for his assistance in designing the figure in chapter 3. I would also like to extend my deepest gratitude to Helene Samantha Dansholm for having corrected errors and inconsistencies in the language, structure and flow of the book manuscript. Without your help, often under time pressure, the readability of the book would surely have been lower. I would also like to thank Alex Giralt García for having fact-checked part of the book and for double-checking that sections related to the hazard phenomenon are based on more solid foundations by cross-checking them with reports from the Colombian Geological Survey. I would also like to thank Gelvin Wilches Herrera for assisting with transcriptions and interpretations of news media. Your help made all the difference.

I would also like to thank JC Gaillard for taking the time to write a foreword for the book. As a pioneer in realising the vulnerability paradigm and an inspiration for myself and so many other emerging scholars in the field your support means a lot.

This book would not have come about without the support and encouragement of Ilan Kelman, series editor for this Routledge book series. Thank you for your time and dedication, as well as your astute inputs. Your encouragement to publish a book in the series is what initiated this whole process to begin with. If there are two people who have inspired me in my career it must be you and JC. Thank you both for lifting up those around you. You are an inspiration for us all.

An expression of gratitude to the publishing team is also in order. A big thanks to Faye Leerink for your enthusiasm for my book proposal and for your support and patience. I would also like to thank Ruth Anderson and Nonita Saha for excellent communication and support throughout the process. Thank you for responding promptly and thoroughly to all my questions.

Part I

Context

1 Introduction

Life in anticipation of slow calamity

> A major feature shared by various creeping environmental problems is that a change in this type of environmental problem is not much worse today than it was yesterday; nor is the rate or degree of change tomorrow likely to be much different than it is today.
>
> Michael H. Glantz (1999: 3)

High up in the Colombian Andes, in the north-easternmost part of the departamento[1] of Nariño, bordering Putumayo and Cauca—opposite the mighty Volcán Galeras,[2] leaving the city of Pasto—lies the resguardo[3] Inga de Aponte. Colombia has experienced one of the world's most protracted guerrilla wars, and although the war formally ended in 2016 after the ratification of a peace treaty between the Government of Colombia and the Fuerzas Armadas Revolucionarias de Colombia (FARC), the peace agreement is still contested and remains in a fragile state. During the peak of the Colombian conflict, the Inga, the indigenous people living in the resguardo of Aponte, succeeded in resisting guerrilla occupation of their ancestral lands. Numbering just under 4,000 inhabitants in a resguardo spanning well over 22,000 hectares, the Inga were awarded the United Nations Development Programme (UNDP)-backed Equator Prize[4] for their nonviolent resistance in 2015:

> Following expulsion of these groups, the Inga people set aside the majority of their land as a 17,500 hectare sacred area. The community organized themselves around a local governance model that is based on a shared vision of justice and collective action on health, education, community services, ecosystem restoration, and sustainable livelihoods. At the same time, the group created the Court of Indigenous Peoples and Authorities from the Colombian Southwest, designed to support other indigenous peoples in reclaiming their ancestral territories and combatting drug trafficking.[5]

This description fits my own observations well. Aponte is governed by a council consisting of various 'ministers', including for health, education and environmental affairs, of which women hold many central posts. Young people are also highly engaged in these decision-making processes, and attendance is not exclusive or restricted. In other words, the Inga in many ways demonstrate many of the qualities

DOI: 10.4324/9780429288135-2

typically associated with well-governed and resilient communities, such as broad participation in local decision-making. Socially and politically, the community is characterised by high levels of inward trust, social cohesion and inclusion in terms of active participation in communal political processes—a rarity in Colombian villages.[6] Power and representation are shared widely, with no single person holding excessive influence over the council or decisions of collective concern in general. Furthermore, one of the best elementary schools in the municipality of El Tablón the Gomez is also located in Aponte.[7] In terms of livelihoods, the Inga produce some of the beans that make up the world-renowned Nariño coffee,[8] providing many of Aponte's coffee growers with a comparatively decent income. However, these seemingly resilient characteristics notwithstanding,[9] all is not well in the otherwise idyllic Andean town.

According to the community governor, Janamejoy, the slow-onset hazard affecting Aponte began to manifest early 2015 after many residences had gradually and mysteriously crumbled due to what initially seemed to be a one-off geological event. Well over six months later, hundreds of residences were already cracking up or crumbling, affecting more than 400 individuals. This is when it became clear that the hazard was not an event but a process—or rather, when it became clear that the individual geological episodes were not isolated events, but were, rather, connected occurrences attributable to the same phenomenon. The hillside was in motion—collapsing day by day, week by week, month by month. In other words, impending calamity was now a reality for the Inga of Aponte, who, from that moment onwards, have lived in anticipation of the eventual but certain destruction of much of their ancestral territory.

Disasters differ markedly in terms of both their manifestation pattern and speed of onset.[10] The nature of their onset, in turn, largely determines how agencies and academics interpret and react to them. Disaster assessments reveal that the adverse impacts of droughts, El Niño warm events and other gradually manifesting disasters constitute a significant proportion of the global disaster burden. Largely ignored by disaster scholars, gradual forms of destruction have been defined away as phenomena not constituting disasters—perhaps mainly for the sake of clarity and comparative reliability. As a result, few theoretical analyses in the field have devoted significant attention to the 'onset' phase of disaster: a phase largely characterised by anticipation of their often-impending, although by no means inevitable destructive effects. This book represents an attempt to develop a better understanding of what happens throughout the manifestation or onset phase of a disaster. It is inspired by the observation that few disaster scholars have previously looked at and theorised the onset phase of a disaster, which, in turn, is probably connected to the very short onset phase of the most frequently investigated disaster occurrences. For example, it would have been logistically and methodologically challenging to design research on how the inhabitants of New Orleans experienced the days and hours prior to the arrival of Hurricane Katrina without drawing solely on retrospection, to say nothing of the difficulties associated with carrying out such research in more remote areas. Hazards whose adverse impacts manifest slowly, conceptualised as slow calamities throughout this book, are better suited for generating insights into how communities experience disasters during their manifestation (or onset) phase.

Intuitively we may think that disasters with a creeping and incremental onset should be easier to manage than more rapid and unforeseen ones. Slowly emerging disasters provide not only a longer period of warning signals, but also a longer period in which proactive steps can be taken to minimise their negative impacts. In reality, however, timely intervention in slow calamities usually fails to materialise due to a number of organisational, psychological, political and economic factors. Most of these factors are connected to the perceived lack of acuteness associated with such processes—rendering them ignorable in competition· with seemingly more pressing societal challenges. Meanwhile, however, their destructive potential may gradually build up and intensify as time passes, until their low-grade, gradual impacts eventually manifest as a full-blown emergency for affected populations and local authorities.

My purpose in this book is to extend disaster theory beyond its present focus on disaster precursors and aftermaths[11] by analysing the 'lifespan' of a slow calamity, by which I mean the period from when the adverse impacts of the hazardous process first emerged until the time after which the place in question was nearly completely destroyed, in this case a period of about four years. I do so by bringing attention to, and providing an account of, the gradual onset of a disaster and the subjective experience of living through the gradual onset of a slow calamity. The title of this book, *Living in Anticipation of Slow Calamity*, signals that in the following chapters I will provide an account not only of slow calamities and their consequences, but also of life during an onset—or life in anticipation of (impending) disaster. Drawing on fieldwork and secondary sources from Nariño in the Colombian Andes, this book focuses on local narratives and conceptions of the gradually unfolding disaster process developing in Aponte, a small indigenous town that faces imminent destruction from a slow-onset geological hazard—a calamity still unfolding as this book project was initiated (see Chapters 2 and 3 for more on the hazard and its adverse consequences). This book is not primarily intended as an historical or geopolitical account of the Aponte situation, but as a contribution to disaster scholarship drawing on this case. This particular disaster process not only provides important insight on the nature and phenomenon of disaster, but also reveals a number of theoretical and conceptual dilemmas and paradoxes stemming from the more 'sudden-onset' research contexts on which many of the frameworks and perspectives used in the field of disaster research are based.

1.1 Key literatures and terms engaged with in this book

Terminology has been a subject of considerable debate within disaster research.[12] Not only do vocabularies vary between languages and policy contexts, but concepts are also interpreted differently across academic schools of thought. Variation in the meaning and interpretation of such concepts as *disaster* or *resilience* is both a source of controversy and a paradigmatic innovation in the field. It may be helpful, therefore, to clarify a few of these controversies and render explicit the theoretical basis of the book. The following paragraphs thus introduce the key terms and themes with which I will engage in this book.

1.1.1 *The nature of disaster*

Reducing the global disaster burden by shifting from a response-focused disaster management strategy to a more vulnerability and disaster risk-centred approach is in many ways a precondition for sustainable development. Societies around the world differ markedly in their ability, willingness and strategies to deal with future hazards, both known and unknown in the present. Differences in economic prosperity, prevailing political constellations and cultural attitudes towards risk, among other factors, largely determine societies' susceptibility to disaster. It is therefore useful to consider the disastrous impacts that hazards often inflict on exposed and vulnerable communities as a product of society itself—not of nature.

Inquiries into the root cause of disasters trace these human tragedies back to underlying and unaddressed conditions of vulnerability that have often accumulated over decades and centuries. Such disaster risk creation can take many forms, ranging from ideologies favouring unchecked natural resource extraction and economic exploitation to inadequate enforcement of building codes or contingency planning. Insights from such analyses therefore reveal that hazards do not produce disasters in and of themselves—vulnerability does. Such an anti-deterministic stance, in turn, shifts focus away from narratives portraying disasters as natural, one-off events, instead rendering disasters inherently political phenomena by pointing out the complex ways in which intersectionalities and structural pressures converge to produce the adverse impacts often (erroneously) attributed to hazards. In short, disasters do not simply happen, but are woven into the very fabric of everyday life, socio-political dynamics and actions—the consequences of which are rendered explicit upon the occurrence of potent natural hazards.

From a vulnerability perspective, the vast material destruction and human suffering produced by hazards may be more productively framed as symptoms, rather than the cause of disaster. From this viewpoint, disastrous impacts are more a result of long-term processes of vulnerability creation than of the hazard itself. Recent developments within this approach have also questioned conceptualisations that define disaster as the adverse impacts of hazards, seeing instead the continued existence and perpetuation of vulnerable conditions as what constitutes disaster. By rejecting the term *natural disaster*, and therefore challenging deterministic explanations for disaster impacts, vulnerability scholars have increasingly recognised that there is no such thing as a rapid-onset disaster—rather, all disasters are, by definition, slow onset. What does this imply for the title and topic of this book, then?

Admittedly, this conceptual dilemma has proved difficult to overcome in discussing the adverse impacts of (and disasters associated with) slow-onset hazards. In writing this book I have chosen to employ a disaster conceptualisation where I take the terms disaster and calamity to refer to the adverse consequences of hazards, as opposed to the view that the existence of vulnerable conditions is what constitutes the disaster by itself. While this does not mean that I do not endorse this view, I believe that such a conceptualisation does not allow a fruitful discussion of the onset phase of a disaster (which differs based on its duration and manifestation dynamics), which is the topic of this research effort. Hence, rejecting the very term slow-onset disaster arguably forces us to adopt a less straightforward vocabulary, meaning that we would have to label such processes, for instance, 'a study of the adverse consequences associated with slow-onset hazards', which I argue would be a clumsy way to conceptualise the subject matter. While recognising that all disasters are a product of gradual process of vulnerability creation, we may still need to maintain a term that refers to disasters associated with slow-onset hazards. The term *slow-onset disaster* or slow calamity or whichever concept we opt for can in this way provide contrast with disasters whose adverse impacts are felt and experienced in more immediate ways, such as the lived experience of an earthquake disaster, despite their root causes being the result of equally long-term processes of vulnerability creation.

1.1.2 *Community*

Disaster scholars have for some time argued that disasters always occur in specific localities and are ultimately suffered locally, regardless of the disaster's overall geographical scope and the severity of its impacts. In other words, disasters are always experienced by households and smaller units of governance one way or another regardless of how many localities are affected. Communities serve as one of the key geographical units of analysis employed by disaster researchers and also as the focus of this book. Furthermore, the quality of communities, the social ties within them and the way they are governed are also among the factors that are assumed to positively influence the ability of communities to cope with hazards and mitigate their adverse consequences. However, the popularity of employing a community lens in analyses of disasters and vulnerability to them has also raised critical questions concerning the legitimacy of naïve conceptualisations of communities and their dynamics. For example, in a seminal paper on the concept, the disaster scholars Alexandra Titz, Terry Cannon and Fred Krüger identify several challenges connected to the concept of community.[13]

First, the concept 'is used "in a bewildering variety of ways", very often without reflection on its meaning, or even a definition'.[14] It therefore often remains unclear in many works whether references to community denotes a people, a physical place or a non-material (e.g., digital) community. While I will not stipulate a concrete definition of community that will be employed consistently throughout this book, the following description serves the discussions that will follow well, although with a few minor points I will clarify below:

> The idea of community comprises groups of actors (e.g. individuals, organizations, businesses) which share a common identity or interest. Communities can have a spatial expression with geographic boundaries and a common identify or shared fate. [Community may be seen as] a place-based concept (e.g. inhabitants of a flooded neighbourhood), as a virtual and communicative community within a spatially extended network (e.g. membership of crisis management in a region) and/or as an imagined community of individuals who may never have contact with each other but who share an identity or interest.[15]

Throughout this book, references to *community* logically refers to the people associated with Aponte, the community introduced at the start of this chapter and whose characteristics and story are the topic of the next chapter. However, according to the definitions above and the critiques raised in the literature, such a delimitation is not as simple as it may sound. At the most basic level, at least two questions remain unanswered.

1.   Is the community geographic (the Aponte area) or ethnic (the Inga)?

> The first question is connected to whether a community is geographically bound to a specific location, and raises important issues connected to resettlement, like how the community can, to as great an extent as possible, be reconstructed at an alternative site. It also forces us to think about the people who have already left Aponte for cities like Bogotá or Cali and whether they are still part of the community, given that they may or may not return or maintain relations with the inhabitants of Aponte.

2.   Is everyone who resides in Aponte part of the community (farmers and other people who have moved there, marginalised groups etc.)?

> The second question concerns the rights and group-membership of non-Inga residing in the Aponte reserve. From a purely geographical conception of community, *campesinos* and local entrepreneurs who have moved to the reserve make up part of the community as well. Socially, there are also obviously interactions and ties between Inga and non-Inga residents. On the other hand, considering the rights of indigenous peoples and the resettlement process itself, the rights of non-Inga are less well defined. Hence, asking the question of what constitutes community membership and what does not, forces us to ponder difficult questions about indigenous status, ethnicity and social cohesion. Of course, some Inga and non-Inga alike may ultimately decide to not follow the resettlement plan and instead migrate elsewhere. How easy it will be for people who have left to return at a later stage to what we may label 'new Aponte' also remains an open question that will have to be explored in future studies.

Second, the concept of community is often used as 'a "moral licence" that supposedly guarantees that the actions being taken are genuinely people-centred and ethically justified'.[16] Given its tendency to generate positive sentiments of inclusion, equity and grass-rootedness, evoking the community term may signal attempts at legitimising certain practices—thus implying that interventions or studies are necessarily 'good' (given their community-centredness); additionally, the term masks questions of social cohesion, internal power struggles and potential discriminatory practices. As such, while employing the term it is useful to have addressed 'the question of who these "people" are' and the degree to which 'they represent a consistent body of individuals sharing the same ideas, perceptions and interests'.[17] Accounts involving community experience must therefore render explicit the inherent tensions within communities as well, rather than gloss over these as observations that do not fit neatly within how we otherwise understand communities as coherent entities.

Third, as suggested by the above, communities are fuzzy—geographically, socially and politically speaking. One example of this fuzziness is the existence of diasporas living as parts of emigrated communities. These diasporas may or may not maintain ties of varying strength to the community, but they may also be considered as communities in and of themselves (such as references to e.g. the 'Filipino community'). We have also seen that a community may consist of sub-communities or cultures. In Aponte, for example, the Inga have coexisted with non-Inga for decades. Additionally, among the Inga there are also hierarchies involving various titles, such as the chief, among other positions of authority. For these reasons it makes sense to see communities not as clear-cut and well-defined entities, but as a relatively fuzzy term.

Fourth, communities are dynamic and change over time; they are not static entities. According to anthropologist Anthony Oliver-Smith, 'communities do not construct themselves—they evolve'.[18] A community cannot simply be uprooted and built up again elsewhere; it will have to be formed again. The question thus arises whether it is then the same community at a new location, or just a new community and that the old community no longer exists. If we see community as a relational concept, it makes sense to consider community resettlement as achievable. Yet, recognising that communities are also dynamic, we would expect the community to change, often considerably, following resettlement and the subsequent process of community rehabilitation precisely due to the relational nature of communities.

Lastly, and of particular relevance to the following section, a community may be characterised by relatively high degrees of social cohesion among its members, while simultaneously being marginalised and/or being suspicious of society at large. While it is important to recognise that communities are always embedded within societies—interacting also with other communities and operating under the jurisdiction of offices at varying levels of governance in the nation-states in which they are situated—these relationships may be characterised by varying degrees of conflict. Put differently, as stated earlier in this section the degree of tension within communities varies, but so too does the degree of conflict between the interests of a community and the wider society in which it is embedded. Communities may therefore, to a lesser or greater extent, be trustful or distrustful towards neighbouring groups and local, regional and national institutions.

1.1.3 Dilemmas arising from distrust and the principle of subsidiarity

Trust and high degrees of social cohesion facilitate coping with— and recovery from—disasters.[19] Conceptually, we can see trust as operating at different levels of abstraction. For example, trust at the micro level relates to trust between individuals, or an individual's propensity to trust, whereas meso- and macro-level conceptions of trust revolve around trust-facilitating institutions and the societal tendency to trust, or the 'trust climate' in general, respectively.[20] Similarly, the causes of trust and mistrust operate at various levels, ranging from individual experiences to longstanding patterns of structural discrimination and conflict.

Colombian society is largely shaped by a climate of distrust, which is arguably a result of its decades-long violent conflict and high levels of class division, crime and homicide.[21] Within this wider climate of distrust, individual communities in which social cohesion is higher, such as among the Inga, social ties and trust levels can be significantly higher than they are in society at large. It therefore makes sense to say that the Inga demonstrate high levels of inward social cohesion, but distrust towards the wider society in which their community is embedded. This introduces a dilemma for the resettlement process in light of the so-called principle of subsidiarity, or the *principle that disasters should always be managed at the lowest possible level of governance* (normally the local level),[22] in this case by the office of the Governor of Pasto, the departmental or subnational authorities. However, a long history of conflict between indigenous communities and local as well as regional authorities makes such cooperation fraught.

While researchers in the field rarely disagree on the fact that communities need to play an active role in reconstruction and resettlement efforts,[23] there have been few discussions of how to address the problem of actors and leadership. During a town meeting in Aponte, it became clear that other actors than the local-level authorities would have been preferred by some leaders either in a leading or supporting role. Actors mentioned included agencies of the United Nations, other international organisations and non-governmental organisations and agencies of foreign countries. International actors are frequently involved by the Inga as distrust towards regional and national authorities is low. It is therefore clear that the principle of subsidiarity may be at odds with other normative ideals, such as autonomy and being able to choose the kind and type of assistance one desires—a dilemma grappled with throughout this book.

1.1.4 Displacement, resettlement and recovery

While this is a book about the experience of living in anticipation of a menacing calamity and not a book on resettlement per se, a brief contextualisation and conceptualisation of the displacement and resettlement process is in order. The related concepts of displacement, resettlement and recovery are of particular interest for us, and I see their connection as follows.

Displacement occurs for a wide variety of reasons, including disasters, conflict, wars, environmental change, large-scale development projects (like hydroelectric dams) or other disruptions, involving inhabited places becoming partially or fully uninhabitable, either objectively or subjectively speaking (in the sense that a place may either be virtually impossible to inhabit due to the complete or partial destruction of a settlement that may or may not be feasible to rebuild or a sense that a settlement is currently too risky to inhabit). While efforts should be made to assist communities in recovering *in situ*, recovery occasionally has to take place at an alternative site, such as in Aponte where most of the original site was rendered too exposed to further disaster risk as the hillside kept slowly moving, potentially culminating in a sudden and dramatic collapse.

Although resettlement is an option that should be considered a last resort, it can also be thought of as an attempt at recovery at an alternative site, when the current one may be deemed excessively exposed to disaster risk. Such a conceptualisation is useful because it not only forces us to think about recovery success, thereby bringing to the fore the challenges of successful resettlement, but also because it reminds us that resettlement is not merely about rebuilding any number of standardised houses at an alternative, presumably safer location, but a longer process of healing and reconciliation. In this process, however, the dynamic and continuously evolving nature of communities should be recognised. As such, reconstructed and resettled communities cannot simply be preserved; they change to a lesser or greater extent, but the quality and inclusiveness of the resettlement process and the way it is governed is assumed to greatly influence success. Failure, on the other hand, may at worst lead to community collapse and at best to escalation of pre-existing conflicts and to persistent, lower, socio-economic outcomes. In the case of Aponte, the ultimate outcome of the resettlement process is not known at the time of writing, but the stakes are clearly high.

1.2 The aim of this book

The main ambition of this book is conceptual advancement, with inferences being exemplified and illustrated by my account of the Aponte disaster. A more descriptive treatment of the case, and in many ways a precursor to the present work, has been provided elsewhere.[24] In saying that my main motivation is to advance theory, however, I do not mean to imply that it is not practical or relevant for policy. My motivation for underlining the theoretical ambition of this book is to make clear that the book is more about disasters and disaster as a phenomenon than it is about the particular series of events that occurred in Aponte. In other words, my unit of analysis throughout the book is not Aponte per se, but calamities that manifest (or the impacts of which present themselves) gradually. Concisely put, the book grapples with the following theoretical puzzles which I believe to be of interest to disaster researchers in general.

1.2.1 *Conceptualising slow and oftentimes elusive manifestations of disaster impacts*

While in writing this book, I do not set out to define slow calamities as such, one of my principal motivations is to spark new debates on how the term should be understood. As previously mentioned (see the previous section), vulnerability theorists commonly argue that there is no such thing as a slow-onset disaster. Surely there are alternative concepts that may be used, such as 'disasters associated with slow-onset hazards', and resolving this debate once and for all is not my main goal here. I believe vulnerability scholars are correct to point out that the term slow-onset disaster to some degree distracts us from established views on disaster causation, emphasising the historical build-up of vulnerability as the root cause of their destructive potential. This is perhaps not how the term would be understood in ordinary usage. As some readers may already have noticed, in writing this book I have avoided making a firm decision on terminology, although I will mainly employ the term slow calamity, instead opting for a flexible writing style where I also write, for instance, 'disasters whose impacts manifest gradually', 'the adverse impacts of slow-onset hazards' or 'disasters associated with slow-onset hazards'. In so doing, I hope to make it clear that I am not dogmatic in my thinking on how we should label these phenomena and people's experience of them. My main goal is to facilitate further debate on how disaster researchers can conceptualise such events in a manner that is accessible to people outside the field, including practitioners, without degrading its terminological coherence, and in particular to theorize how these phenomena are to a significant extent shaped by 'living in anticipation'.

If we accept the notion that all disasters are slow-onset, there is potentially nothing that distinguishes the Aponte case from similar communities destroyed by other kinds of disasters, such as disasters associated with earthquake hazards. One obvious objection would be that because vulnerability is the cause of both examples, theoretically nothing really separates the Aponte case from a study of similar places affected by so-called rapid-onset 'events'. As I have previously made clear, I do not object to this theory of disaster causation; indeed, I employ a vulnerability framing in this book and the rest of my work. At the same time, the subjective experience and the real practical challenge associated with gradual destruction is theoretically and pragmatically affected by hazard onset speed. In other words, in the eyes of community members and practitioners, as well as political leaders, a hazard whose impacts unfolds over a four-year period surely differs from one that triggers a disaster in a matter of minutes without forewarning. As such, a starting assumption that gave rise to this book, therefore, is that disaster scholars may have prematurely concluded that distinguishing between rapid-onset and slow-onset disasters is of little practical and theoretical significance. While I realise that this claim may be controversial, I would consider it a great success if this book inspired efforts to engage in this debate so as to clarify not only how we should label such phenomena (forms of destruction and plights suffered slowly), but also disaster theory in general.

1.2.2 What bearing does disaster onset speed have on the subjective experience of disasters from the perspective of affected communities?

Disasters take place at varying temporal scales, and as we have already seen they are often productively understood as incubating phenomena that are woven into the very fabric of society through complex historical processes of disaster risk creation. An alternative perspective is to see the onset of disaster as the point at which the adverse consequences of a hazard become visible to the public and are experienced by affected populations. Such a perspective may be labelled the subjective experience of disaster, and reminds us that the notion of disaster as a totalising phenomenon may be approached through a whole array of perspectives, ranging from historical, political and construction to phenomenology or constructivism, or a combination of these.

Analysing and theorising onset speed as a central unit of analysis in a disaster setting from the perspective of subjective experience of affected households adds to the existing literature by providing a better understanding of how the anxieties triggered by gradual, perpetual types of destruction are experienced, and, indeed, lived with. Here it is important to reiterate that life in anticipation of slow disaster is characterised simultaneously by continuity—in the sense that daily life still has to be lived, but also by a melancholic sentiment of impending displacement and its resulting uncertainty. The gradual manifestation of disaster impacts is in this way experienced differently from sudden, unambiguous and immediately visible forms of destruction. Not only is there an absence of sentiments of urgency, drama and immediate risk of personal survival, but the ambiguity of the onset phase means that it may for some time be unclear whether the destructive process will be linear or exponential, temporary or permanent, or preventable or unstoppable, all of which shape affective aspects of the disaster.

1.2.3 The displacement dynamics of slow calamities

By some estimates, disasters now displace a greater number of people than wars and conflicts.[25] Reports suggest that just under 30 million people are displaced by disasters annually, the majority of whom become internally displaced—or displaced within their own countries. While disaster-induced displacement is better understood than previously due to increasing research interest, the way displacement works and affects communities remains insufficiently explored compared to many other central topics within the field of disaster studies. Additionally, while studies of droughts as well as environmental and climatic change have provided more impetus to understand how displacement functions in the context of slow-onset phenomena, the particular displacement dynamic of slow calamities remain undertheorised.

In some cases, distinguishing between internal displacement and people who have simply relocated for unclear reasons can be difficult. Nowhere is this more evident than in the case of protracted and ambiguous forms of hardship attributable to a myriad of factors rather than a single triggering event. Ethnographic evidence suggests that households rarely take the decision to uproot their lives and relocate light-heartedly, and mainly do so after their coping capacities have been exhausted. Drought-induced displacement, for example, may be triggered by a combination of factors, like inadequate recovery from previous drought episodes, lack of institutionalised safety nets, lack of access to reliable information or forecasts converging with external interests. Thus, like disasters, therefore, displacement is rarely caused solely by the existence of a natural hazard. Rather, it is a result of longer processes of discouragement culminating in a decision to uproot and settle elsewhere, temporarily or permanently. The link is particularly ambiguous in the case of slow-onset disasters, where the point at which households are overwhelmed is better understood in terms of some threshold eventually being reached, as opposed to a more instantaneous phenomenon that caused material damage which rendered homes uninhabitable overnight.

The ambiguous nature of the adverse impacts associated with most slow-onset phenomena means that causal inferences may often largely be characterised by subjective attribution. When examining displacement in Aponte, there are methodologically challenging decisions about attributive factors to consider. For example, there exists the question of whether to assume that all the inhabitants who have moved since 2015 were displaced as a result of the slow calamity, or whether some may have moved for entirely unrelated reasons. In reflecting on the displacement dynamics of slow-onset phenomena, as well as other themes, I will refer to what I call *attribution problems*, or reflections on causality and causation claims more generally.

1.2.4 How can we understand resilience in the context of slowly but perpetually worsening disaster impacts associated with a particular hazard phenomenon?

The popular notion of (disaster) resilience has been described in the literature as a metaphor, theory, set of capacities and a strategy for disaster preparedness.[26] Taken from the fields of mechanics and ecology, the resilience term has been applied metaphorically in disaster studies, where it has, in broad terms, been taken to mean the ability of individuals, communities or societies to resist, bounce back from and become less vulnerable to disasters. As a theory, resilience has often been seen as an antonym or solution to vulnerability, stressing the importance of social capital, adaptability and enhancing coping capacities.

Two important but overlooked questions within the literature on community disaster resilience are: first, is it even possible to be resilient to all sorts of hazards?; and second, can a community still be considered resilient if they were forced to resettle due to the existence of a hazard whose impacts will not stop and whose impacts will worsen indefinitely? In other words, does resilience as a metaphor or theoretical concept make sense in the context of 'unstoppable' slow-onset hazard, and can resettlement be considered a resilient move or only as a proof of lack of resilience?

1.2.5 How do distrust and dilemmas connected to the principle of subsidiarity affect community involvement in resettlement processes?

Another puzzle I touch upon in this book is the dilemma between community involvement in the resettlement process, or rather the autonomy of communities in deciding which actors should be involved in the process, considering the principle of subsidiarity. The principle of subsidiarity is one of the various key principles of disaster management underpinning policy and practice in many national and subnational systems in the world, although each system may institutionalise them differently:

1. *The principle of responsibility*
    This principle stipulates that the agency or actor responsible for one policy area or function has the main responsibility for carrying out contingency planning and service delivery during disaster situations. Its intention is to ensure accountability and division of labour, and to ensure that the most relevant or

competent actor within a policy domain is responsible for contingencies relating to that domain.

2. *The principle of similarity*

   The similarity principle entails that the organisational structure and operations during disasters should resemble, as much as possible, its normal functioning or modus operandi. The principle is based on general insights from disaster research noting that ad hoc re-organisation as well as re-shuffling of staff, hierarchies and functions further complicates coordination and the management of disasters.

3. *The principle of subsidiarity*

   Constituting in essence a normative goal in favour of decentralisation, the principle of subsidiarity stipulates that disaster risk and disasters should be managed at the lowest possible level of governance. The salient point is that the actors closest to the disaster area are normally better equipped to understand the local context and get an overview of the situation and are therefore better capable of managing it.

These most basic and universal principles are also parsimonious with principles identified as best practice in the research literature. These are also often followed by one or more additional principles, such as principles requiring coordination with relevant actors, principles stressing the autonomy of subnational federal systems or principles underlining an 'all-hazards' approach to disaster planning.

A decentralisation logic, such as the one underpinning the principle of subsidiarity, implies that the mandated responsibility to assist communities unable to hold their own falls on the local or subnational authorities. In the case of Aponte, their own *cabildo*, or community council, lacks the capacity and resources to carry out the resettlement and recovery process by itself. This responsibility, therefore, falls on the authorities in the departmental capital of Pasto. However, during community meetings long-standing patterns of discrimination and regional politics became evident, which have caused communities in and around the municipality of El Tablón de Gomez to question the degree to which the subnational authorities in Pasto may be trusted to have their best interest in mind. Since the chain of mandated responsibility is clear from the perspective of the principle of subsidiarity, communities will generally have to accept assistance from the authorities legally mandated to provide it. At the same time, communities may have other relevant actors in mind, but whose intervention and assistance would ultimately put them in conflict with the authorities officially in charge of the process. A dilemma arises, therefore, in which communities may want the assistance of national authorities, neighbouring authorities, non-governmental organisations (NGOs), UN organisations or foreign agencies, but whose involvement may contradict the principle of subsidiarity. This book reflects on this dilemma from multiple perspectives, most of which are only loosely related to the actual Aponte case but still inspired by it. My goal is that these reflections, even the ones not directly derived from the situation in Aponte, will be useful to researchers operating in other contexts and those who have a general interest in questions of community participation, decentralisation and conflict.

1.3 Living in anticipation of calamity

As the title of this book suggests, the main theme in this book is the analytical notion of living or life in anticipation of calamity, a theme that touches upon all the puzzles identified above. The analytical frame of life in anticipation of disaster is ideal both for studying slow-onset disasters, such as the one that occurred in Aponte, as well as projected/expected disasters that we, through forecasts or technical surveys, know will affect a particular community in the future—a sort of sentiment of 'we live here for now, but perhaps not for long'. In other words, the notion is based on the idea that after a community is made aware that their future existence is threatened, either by a current slow-onset process or by a forecast future event, their continued existence in said location is characterised and indeed shaped by a sentiment (to a lesser or greater extent) of living in anticipation of disaster. In this subchapter, I will focus on the analytical notion in general, both as it relates to the Aponte case and in terms of how it may be used in other studies where the concept may be usefully applied.

Applying the concept in slow-onset disaster contexts, such as in Aponte, not only makes possible studies of how people make sense of slow changes as they occur, but also of how daily life inevitably and necessarily continues throughout the onset period. Applied to the study of slow calamities, the notion of life in anticipation of calamity is uniquely suitable for studying, in particular, the (slow-) onset phase of gradual destructive processes, paying particular attention to the dualism of disruption and continuity, as well as the feelings of anxiety that a life in anticipation inevitably evokes. This book pays attention to both aspects of the concept identified here.

First, and concerning gradually occurring calamities, how people make sense of such phenomena naturally depends on a number of factors, including culture, risk perception, education and socio-economic standing, among other factors. Additional concerns and key determinants of disaster experience include hazard characteristic and the way in which its (potentially) disastrous impacts manifest in terms of loss of life, injury, material destruction, disruption and otherwise. Disaster onset speed can, in this way, be understood not only as a central component of the lived experience of affected communities, but also how they make sense of, or indeed live with its gradually intensifying consequences for a prolonged period of time. Because the consequences of slow-onset disasters are suffered gradually over time, they are likely to be experienced less dramatically and overwhelmingly than more rapid calamities, with their typically immediate and hugely destructive consequences unleashed all at once. Hence, the manifestation pattern of disasters affects how they are conceptualised by affected populations and how their adverse impacts are suffered and coped with.

The realisation that a gradual change process is underway starts with a series of observations that, taken together, appear to be part of the same process or phenomenon, as opposed to being singular, discrete events (indeed, disaster researchers often see disasters in general as processes, or symptoms of systemic vulnerabilities rather than as isolated, disconnected events).[27] Accordingly, when discussing slow calamities, it may not initially be obvious whether discrete symptoms, like the cracks that appeared in the foundations of several homes in Aponte in 2015, are just a few cracks or part of a larger phenomenon that will worsen exponentially over time. Similarly, until such phenomena are properly surveyed, determining whether the underlying hazard phenomenon is about to dissipate, or whether it will go on causing harm indefinitely, rendering the place in question uninhabitable may be difficult. In other words, at any point in time affected community members may be asking themselves whether the impacts of today were the last (that the process is now over and recovery activities may be initiated), or whether the slow-onset disaster will ever pass (triggering a sentiment of 'we cannot remain here for much longer'). These questions and anxieties are all part of the experiential aspect of life in anticipation of disaster.

Life in anticipation of disaster is shaped both by continuity and anxiety, and invites analytical attention to sentiments connected to the eventual (but not immediate) destruction of a community and how these evolve throughout the onset phase as economic, social and political life inevitably continues if the onset phase is of a long duration. While the realisation that one's livelihood, land tenure rights, home and family ties and community ties are all threatened is a major cause of anxiety and melancholia (perhaps even existentially, as there is always a risk that these assets may never be fully or even partially recovered in the aftermath of a disaster), life also has to be lived until destruction and/or resettlement is a fact of life. This means that during the onset phase children must attend school, farmers must continue harvesting crops, shopkeepers must continue selling goods and service providers must continue offering their services to their customers. At the same time, faced with the prospect of displacement, identity loss, further impoverishment or even community collapse (following, e.g., an unsuccessful resettlement and recovery effort), the lives of people living in anticipation of disaster can be expected to, first and foremost, be characterised by dread, helplessness, uncertainty and melancholia.

The second aspect concerns what may be labelled scheduled hazards—or a hazard that we know will destroy a place once it (inevitably) occurs. Advances in technology allow us to detect previously undetectable hazards, meaning that several places previously considered to be relatively safe places to live are today known to be doomed, so to speak. Thanks to technology and prediction tools, we are now aware that some places co-exist with potential and even inevitable hazards with hugely destructive potential, effectively dooming exposed settlements in the long term. Applying the analytical notion of life in anticipation of disaster to such communities allows us to analyse both how living in the shadow of inevitable (in the absence of enormous investments in mitigation), eventual destruction shapes life in such threatened communities, and also whether or how everyday practices are focussed on ensuring that the known hazard will not produce a disaster. In other words, how lived experience become subject to what we may refer to as regimes of disaster risk reduction.[28]

Scheduled hazards take on many forms and disaster researchers have already studied several of them, although in ways that are different from the approach envisaged in this book. Places ranging from California cities and towns, expected to be obliterated once the Cascadia subduction zone and the San Andreas Fault produces 'big ones', and urban areas living in the shadow of massive volcanic eruptions to settlements existing in the advent of an impending landslide hazard are all effectively, to a lesser or greater extent, shaped by life in anticipation of disaster. The attention awarded by individuals to scheduled hazards is, as with slow calamities, mediated by a range of factors—including the nature of official communication about the hazard, cultural frames, socio-economic background, media coverage, societal trust and other factors. Oftentimes, the identification of impending hazards is also followed by some kind of warning and evacuation mechanism, the role of which will also shape how life in anticipation of disaster plays out in practice. This experience is, in turn, mediated by how drills and the socio-cultural construction of future disaster is rendered real through anticipatory social practices. However, life in anticipation of disaster may also, depending on circumstances, be shaped by doubts or anxieties over the effectiveness of these measures or whether people themselves or their loved ones will be able to get out of harm's way. After all, putting one's own life and the lives of one's family into the hands of expert strangers require a deep sense of trust in competence and oftentimes technology.

In sum, the analytical notion of life in anticipation of disaster (or calamity) provides a new perspective on how various units of analysis, ranging from individuals to whole communities or urban areas, experience and act out their lives in the shadow of either a disaster slowly unfolding before one's very eyes, or expected future disasters. As a conceptual tool—a thinking hat, so to speak—the notion of living (or life) in anticipation of disaster invites us, in the context of gradual destruction, to pay attention to questions of affect, continuity and disruption during the onset phase of disasters as they unfold. Contrastingly, in the case of scheduled hazards, the notion concentrates attention on how activities, sentiments and social practices are shaped by life in the shadow of what is perceived as impending (and to some extent inevitable) destruction.[29] Practical and theoretical insights based on this notion, moreover, may produce valuable insights of relevance to other contexts, such as communities awaiting impending displacement from large-scale development projects or any other scheduled displacement or resettlement processes. As such, this book sets out not only to critically engage with several key literatures in disaster studies, such as literature on the disaster concept, but also to present an interesting analytical frame for related fields. The volume is structured as follows.

1.4 Themes and structure

The remainder of this book is divided into three main parts. Part 1, Context, concerns the geographical and theoretical context of the study underpinning this book, and consists of two chapters. Chapter 2, 'Aponte' (the chapter following this introductory chapter), serves to provide background on Aponte and its people, the Inga, including their history, the characteristics of their community and present situation, and to elaborate on the wider context, including the Colombian disaster risk reduction (DRR) system. Chapter 3, 'The phenomenon', aims to inform the reader both of the specific hazard affecting Aponte, and also how we may approach it conceptually and theoretically. Hence, the chapter serves primarily to provide an overview of the hazard, but also aims to conceptualise and theorise slow-onset disasters (the gradually unfolding impacts of these hazards) in a broader sense. Chief among the practical challenges discussed as part of this chapter is the distrust the Inga hold towards experts from the municipal and departmental disaster management agencies, which are formally charged with resettling the community to an alternative location.

Part 2, 'Experiencing slow calamity', concerns local narratives and conceptions of the hazard phenomenon and its adverse impacts. Rather than providing an objective geological account of the process affecting Aponte, the principal aim of the book is to capture the sentiment of life in anticipation of disaster, encompassing both how the hazard is interpreted and understood, and how its impacts are felt and, indeed, lived with (recognising that such processes are characterised both by continuity and disruption).

Chapter 4, 'Making sense of the hazard', aims to outline and analyse local understandings of the phenomenon affecting them. Slow-onset disaster impacts often manifest along diffuse temporal and spatial dimensions. Elusive forms of destruction and decay that occur gradually but with no sign of dissipating will necessarily produce different interpretations than do instantly occurring phenomena. For this reason, this chapter focusses on experiences and narratives surrounding the nature, origin and causes of the phenomenon.

Chapter 5, 'Living with a slow calamity', is devoted to conceptions connected to the onset speed of the disaster. Nearly three years have passed since the first cracks started appearing around and in homes around the indigenous reserve of Aponte, and the worst is yet to come—although around 400 homes have already collapsed. Members of the Inga report having initially witnessed cracks appearing in the soil and grass, which have subsequently expanded and multiplied at an exponential rate over time. At the earliest stages of its onset, nobody in the community could have anticipated that the phenomenon would intensify over time. Understanding local descriptions of the onset from its beginning until the present day provides us with new insights not only on the way in which slow-onset hazards manifest, but also on the challenges associated with understanding the scope and extent of what is to follow. Slow-onset hazard impacts may seemingly pass at any moment and it may be challenging to determine whether seemingly isolated incidents are part of a larger phenomenon. As time passes and the impacts worsen, describing such processes by evoking the term *emergency* becomes increasingly normal. Yet, applying such terms to slow-onset hazard impacts can also be challenging, given their often-undramatic manifestation patterns.

Chapter 6, 'The ancestral land', engages with narratives on the importance of territory for the Inga of Aponte, with particular reference to the planned relocation of the community. Because Aponte is legally considered an indigenous territory, the relocation process not only opens up questions of belonging and remaining on the ancestral land, but also legal questions surrounding the sovereignty of the territory and how this can be moved to an alternative site. Therefore, this chapter outlines and analyses local narratives of place-boundedness and territoriality in the context of displacement and resettlement.

Part 3 of the book, 'Reflections', contains its main scholarly contributions as well as its conclusion. As opposed to the account developed in Part 2, the principal aim of Chapter 3 is to build on this work to further reflect on the analytical notion of 'life in anticipation' as well as to draw out the main contributions of the book for practice and theory.

Chapter 7, 'Life in anticipation', reflects on the material presented throughout the book with the purpose of exploring the lived experience of dreading and awaiting both resettlement and potential community collapse. The last and concluding chapter reflects further on issues raised throughout the book and aims to tie some of the loose ends that remain from the previous chapters.

Notes

1 Departmento, Colombian equivalent of province or subnational jurisdiction. This book operates with the following levels: the national level, departmental level, municipal level, community level and individuals/households.
2 See also: Williams and Montaigne (2001), Bruce (2001).
3 Reserve under various forms of semi- or fully autonomous indigenous governance.
4 For more information on the Equator Prize initiative see: https://www.equatorinitia-tive.org.
5 https://www.equatorinitiative.org/2017/05/30/wuasikamas-el-modelo-del-pueblo-inga-en-aponte/.
6 While it is common for *pueblos* (towns or villages) to be tightly knit my perception is that few have as formalised processes of political engagement and participation as what is the case in Aponte.
7 This statement is not official but is rather what rumor has it in the area. Not only locals in Aponte bragged about the quality of the local school; people I spoke with in Pasto also shared this impression.
8 http://narino.cafedecolombia.com/en/narino/el_cafe_de_narino/
9 One of the main points I try to raise in this book is that resilience building, while necessary, may not always be sufficient. Some disasters can simply be too severe to be 'resisted' or 'bounded forward from'. A question that arises is therefore whether resettlement can be considered a resilient move. See: Staupe-Delgado (2019a).
10 Matthewman (2015), Shaluf (2007), Gundel (2005), McConnell (2003), tHart and Boin (2001), among others, describe and theorise various aspects of slow-onset disasters, including manifestation patterns, typologies, unique challenges and characteristics.
11 Here I refer to the tendency of studies to either focus on vulnerabilities prior to an actual occurrence or causal theories connected to underlying conditions that existed prior to the disaster, and on the other hand the tendency to focus on the response or recovery phase in the aftermath of a disaster. In my view these two perspectives can be seen as what separates the disaster studies literature from the disaster management literature, as the former is chiefly concerned with developing a better understanding of disaster as phenomenon and societal problem, while the latter is concerned with their operational management.
12 The topic of terminology and conceptual change within the disaster research field is extensively discussed in Staupe-Delgado (2019b) and Kelman (2018), for discussions on the disaster concept itself see also Perry (2018), Perry and Quarantelli (2005), Rosenthal (1998) in the bibliography.
13 For a critical discussion on the concept of community see: Titz, Cannon and Krüger (2018).
14 Barrett (2015:182), cited in Titz, Cannon and Krüger (2018: 1).
15 Kruse et al. (2017: 2322), citing also Norris et al. (2008: 128) and Mulligan et al. (2016).

16 Titz, Cannon and Krüger (2018: 1).

17 Titz, Cannon and Krüger (2018: 2).

18 Oliver-Smith (2005: 55).

19 See, for example: Aldirch (2012), Tierney and Oliver-Smith (2012), Nakagawa and Shaw (2004).

20 Murphy (2006).

21 See, for example: Burnyeat (2020), Berents and ten Have (2017), Nussio and Oppenheim (2014), Jimeno (2001).

22 Most countries' disaster management institutions operate under some kind of principle similar to the proximity principle as part of efforts in the 20th century to decentralise disaster management practice. Other important principles also exist, for an elaboration see, for example, Kruke (2015).

23 For discussions on best practices of disaster reconstruction, recovery and/or resettlement see: Davis and Alexander (2016), Oliver-Smith and de Sherbinin (2014), Tierney and Oliver-Smith (2012), Perry and Lindell (1997) and Berke, Kartez and Wenger (1993).

24 In a previous publication I wrote about the Aponte disaster and its implications for how we think about community resilience and recovery. See: Staupe-Delgado (2019a).

25 See: https://www.undrr.org/news/disasters-displace-more-people-conflict-and-violence

26 See: Norris et al. (2008). In this book I have attempted to avoid emphasising jargonised terms such as resilience, vulnerability, exposure, capacities, and others, in an effort to employ mainly natural language. My distinction between hazard and disaster is an exception as this distinction is important for the theoretical base this book builds on and contributes to. I have not tried to avoid using these concepts altogether, I have simply attempted to avoid giving jargon a lot of importance. No particular typological difference is inferred between the concepts of calamity and disaster either.

27 Numerous discussions of the disaster term emphasise the processual nature of the phenomenon, including Kelman (2019a), Bankoff (2004a), Rosenthal (1998) and Oliver-Smith (1996).

28 Referred to elsewhere as 'regimes of anticipation', see: Adams, Murphy and Clarke (2009).

29 I do not mean to suggest in any way that disasters are ever inevitable per se. My point is simply that some areas are exposed to unmitigated disaster risk to an extent where disaster is virtually a given (in the absence of enormous investments in mitigation).

2 Aponte

Political, geographical and community context

This book is inspired by a field expedition I was invited to participate in during fieldwork for my PhD on El Niño preparedness in 2016. My dissertation was, like this book, centred on the onset phase of a disaster, but specifically from a disaster management perspective.[1] At the time my research was focussed on proactive actions undertaken by disaster management actors working in and around the city of Pasto (see Figure 1) in anticipation of the very strong El Niño warm event of 2015–16.[2] A goal of mine was to develop a better understanding of what was done to cushion its adverse impacts by studying precautionary activities undertaken from the time it was forecast until the time it peaked. As part of this work, I was in regular contact with the United Nations Development Programme (UNDP) office in Nariño, where I was also offered office space. I was since given the opportunity to participate in an expedition to an unrelated slow-onset disaster of a geological nature taking place in a town called Aponte. The material for this book is based on the observations, conversations and impressions from this expedition and future meetings concerning the disaster. Moreover, since the onset phase of the disaster lasted longer than my PhD fieldwork did, I have also supplemented my own direct experiences with a variety of secondary data sources, including newspaper articles, archival data, official reports, news coverage and TV/radio interviews with local leaders available online, all of which are cited when used.

Figure 1 Map of the research site and context[3].

DOI: 10.4324/9780429288135-3

2.1 Locating Aponte geographically

My engagement with this story began in Pasto in the spring of 2016. I first heard about the disaster affecting the indigenous reserve of Aponte at both the Gobernación de Nariño and UNDP. The community is located on the borders of the neighboring departments of Cauca (where city Popayán[4] is located) and Putumayo (the site of the more recent and better-known 2017 Mocoa landslide disaster.)[5] I was since invited by UNDP to partake in an expedition to the site and this book in many ways sums up my reflections after having following this disaster for over three years.

The Nariño department is located in the south-westernmost part of Colombia with borders to Ecuador and the Pacific Ocean and consists of 64 municipalities. In the Colombian context, Nariño is considered to be among the main coca-growing and -processing region in the country and is therefore a region with widespread violence and illicit economic activities.[6] As an illustration, during my six years of regular visits to the departmental capital city of Pasto it has never been considered safe or feasible to visit the Pacific coast due to the risk of land pirates as well as violence in the Pacific coastal town of Tumaco. While its average socio-economic standing is comparable to the upper tier of medium-score Human Development Index (HDI) countries, such as The Philippines, there are pockets of considerable poverty, conflict and unemployment in the population living in neglected rural areas and smaller towns.

The drive from Pasto to Aponte takes about two and a half hours, passing through breathtaking Andean mountain landscapes, including the noteworthy canyon of Janambú, and the smaller town of Buesaco as well as the municipal capital of El Tablón de Gómez. The neglect of the more remote and rural settlements along the way is immediately apparent, as is the increasing insecurity as one progresses out of Pasto. In fact, officials in Pasto generally advise against driving North of Buesaco,[7] which is located about an hour's drive north of the city of Pasto. To avoid confusing readers with the places mentioned above, here follows a concrete overview of the most important ones to remember:

Bogotá – the capital city of Colombia, is a metropolitan city with a population of about 11 million, and is home to the national branches of national disaster management agencies, such as the *Sistema Nacional de Gestión del Riesgo de Desastres* (UNGRD[8] – National System for Disaster Risk Reduction) as well as the Colombian Geological Survey (SGC).

Pasto – the capital of the department of Nariño, is a provincial hub with a population of about 400,000, and home to the Gobernación de Nariño, UNDP's Nariño-Cauca-Putumayo office, and the local affiliates of the UNGRD and partner agencies with disaster management functions, which are all important actors in the context of this book.

El Tablón de Gómez – is a municipality in the northeastern corner of the Nariño department with a population of about 13,000, and serves as the local context for this book. It is also the name of the town in which the municipal authorities of El Tablón de Gómez (the municipality) are seated. Throughout this book I will refer to the municipality as El Tablón de Gómez and the town as El Tablón, but the differentiation is not essential for any of the salient points raised in the book.

Aponte – is a community, *cabildo* (village government) and indigenous reserve located 12 kilometers east of El Tablón in the municipality of El Tablón de Gómez and serves as the field site for the insights on which this book is based.

This establishes the national, subnational and municipal context as well as key actors in the book, at least in so far as they are relevant for the practical and theoretical points to be made throughout the following chapters. I will now move on to elaborate on the community characteristics of Aponte, including the disaster which displaced its population. In the sections that follow I will attempt to provide a sense of how Aponte was a highly resilient community in many respects, but that it was still unable to withstand or bounce back from the destructive process it was perpetually affected by until its eventual collapse, at least in the material sense of the word.

2.2 The Aponte community and the Inga people

The Nariñense indigenous populations are mainly made up of Awá—who constitute the vast majority, totaling 26,800 members, as well as the Esperara Suaoudara and the Inga, with 4,500 and 3,600 members, respectively—in addition to a range of smaller groups of indigenous peoples with 200 or fewer members.[9] Common to most of these communities is that they are located far from urban centres in some of the most remote municipalities in the least accessible corners of those municipalities. Furthermore, the indigenous peoples of Nariño are spread over three climatic regions. First, on the Pacific plains along the coastline lowland and coastal communities have thrived for centuries. Second, the highland Andean region which makes up the middle part of Nariño is characterised by its somewhat drier, colder climate. Lastly, a more tropical region lays towards the east as the climatic zone transitions towards the east Andean slopes and the Amazon Natural Region.

The Inga of Aponte formally trace their roots back to Carlos Tamavioy, *the taita de taitas* (Father of Fathers) in the early 1700s, but the territory was before this time part of the northernmost frontier of *Chinchaysuyu*, or the Northernmost territory under the reach of the Inca Empire.[10] The Inga thus consider themselves 'the history and living culture of this millenary nation that during its peak period of expansion and organizational strengthening arrived at Nariño'.[11] The formal recognition of the reserve as it appears today, although not identical, is based on Tamavioy's will, stipulating that:

I, Don Carlos Tomavioy, declare as assets that I myself own, some lands known as Tamoabioy, and I appoint as the executor of my will Don Melchor Juajuandioy. I am letting you know that I am leaving these lands called

Tamoabioy, that stretch three leagues long from the top ravine of Guaraca, to the bottom which we call Aponte, which is downstream Juanambú, it is the middle part, thus, the part between Guaraca and Aponte that I leave my children and all my people, which is my will.[12]

The Inga of Aponte hence trace their heritage back to the arrival of chief Carlos Tamavioy, his wife and his three children, and accompanying families, to the territory in the late 1600s,[13] at which time it was connected to Putumayo and nearby urban centers through mountain trials. Since this time, Aponte has remained predominantly an agricultural community made up of both Inga and other peasants. Its current status as a resguardo, however, traces its origin back to 2003, three hundred years following the founding of Aponte, at which time the then current chief, Hernando Chindoy, set in motion a process that would lead to an unprecedented agreement with the Government of Colombia and the passing of Resolution 013 by the Colombian Institute for Agrarian Reform (INCORA), granting them full protection and the legal rights of their territory.[14] The title of resguardo, then, is considered a collective land title issued to indigenous communities for a defined territory that is inalienable, imprescriptible, and immune from seizure.[15] In short, it would take well over 300 years from initial settlement in Aponte until durable legal rights and sovereignty over the territory could be formally secured, demonstrating not only the perseverance of the Inga but also their capabilities. The relative resilience of this particular community is further illustrated by the process that ultimately caused them to be awarded the Equator Prize in 2015, to which we now turn our attention.

2.2.1 Wuasikamas

During the 1990s, the slopes surrounding Aponte became increasingly covered in the deep red color of opium poppies (*Papaver somniferum*), for the purpose of heroin production (which peaked at over 2,500 hectares of cultivation).[16] According to community representatives, the surge in income traceable to the lucrative trade in opium sparked a gold rush of sorts in the territory, which saw its population increase exponentially, from 1,400 to more than 30,000.[17] Before long, the Andean slopes which were once relied on for the cultivation of peas, potatoes, beans and maize were covered predominately by poppies. In the words of Hernando Chindoy:

> in 1991 when they started with 6 hectares located in the mountains, about 6 hours walking from the town. A year later there were already 200 hectares that were taken around the reservation. And in four years, they increased to 1,000. We had 2,500 hectares of opium poppy, which could produce between 2 and 3 tons of heroin each week. The money also came in, up to 4 million dollars a week.[18]

By the early 1990s a host of armed groups were present in the territory, including paramilitaries, the *Ejército de Liberación Nacional* (ELN), the *Autodefensas Unidas de Colombia* (AUC) and the FARC. Poppy growers were forced to pay tribute to the latter two groups. The Aponte region had thus become the primary poppy-producing region, bringing it considerable wealth, but at a price. Multiple homicides were recorded weekly; people were mercilessly shot in their homes as well as in public. Meanwhile, the FARC effectively kept authorities out of the area while waging war on insurgents, including paramilitaries. According to representatives of the community, citing the Single Registry of Victims, there were over 6,000 victims and affected people recorded in El Tablón de Gomez alone, including as many as 120 members of the Inga.[19]

The situation worsened still further with the ratification of Plan Colombia, a bilateral deal between the United States and the Government of Colombia in 1999, with the purpose of combating drug trafficking and guerilla warfare, through among other means the promotion of illicit crop substitution programmes and other social development strategies. Not only did the armed conflict subsequently intensify, but the aerial fumigation of illicit poppy fields killed crops indiscriminately, also threatening local food security in addition to the adverse health effects associated with exposure to the chemicals used in the war on illicit crops. On the process that followed, Hernando Chindoy wrote:

> In that state of affairs, the authorities of the community of Aponte insisted that their people recover their territorial integrity, autonomy, dignity, and, above all, sovereignty over what belonged to them. In 2003 we undertook the fight for the titling of the ancestral territory as *resguardo*, and once this purpose was achieved, manually and voluntarily eradicated poppy cultivation and created a comprehensive life mandate for the survival and permanence of the Inga people trough time and space. We thus consolidated a territory of 22,283 hectares now free of illicit poppy and freed ourselves from fumigation with glyphosate, in addition to the presence of guerrillas, paramilitaries, the police and the army, which, in the contexts of armed conflict and drug trafficking, violated human rights, international humanitarian law and the rights of Mother Earth, in the constant process of violence lived between 1986 to 2006.[20]

In the process of institutional and cultural change, which included the achievement of self-determination and revitalisation of philosophy, language, spirituality and governance, the cabildo, or town council, was strengthened. In addition to a town leader, a position held by Chindoy at the time, branches of minor cabildos were implemented, which respective ministers of health, culture, education, gender, recreation, youth, economics and services, organised around the principles of *mana sisai* (do not steal), *mana llullai* (do not lie), *mana killai* (do not be lazy) and *alli kai* (be worthy). Underpinning these principles is the notion that 'The Inga man or woman directs their own path and strives to know how to walk according to the teachings that have been inherited from their ancestors, and in their self-determination choose the principles of their life path, facing the consequences in each of decision as individual, family and collective.'[21] This is the essence of the concept of *Wuasikamas*—guardians of the land and keepers of the territory.

The Wuasikamas movement is borne out of the process of cultural revitalisation that followed the political and institutional mobilisation and renewal process, ultimately resulting in the eradication of illicit crops and territorial consolidation. This process was partially financed through an innovative alliance struck with the Government of Colombia through a programme emphasising subsidies for the eradication of illicit crops and subsequent crop substitution. Capitalising on the process, the Inga organised to have a communal fund set up to support community-level efforts at revitalisation and re-development as part of their efforts to keep combatants out and restore their degraded environment. Under the aegis of the Wuasikamas movement, the Inga thus re-organised under a governance model centered on a shared vision of equity and collective spirit under a reformed cabildo model. In 2011 the movement and its leader were recognised as being among the most influential forces for progress and social development in Colombia. The model since spread through the establishments of entities such as the Tribunal of Indigenous Peoples and Authorities of the Colombian Southwest and other legal institutions and processes initiated in the spirit of Wuasikamas, such as Alliance of Indigenous Women of Nariño, for which the community and the Wuasikamas movement was awarded the Equator Prize in 2015.[22]

The concept of Wuasikamas has since been formalised through the registration of a trade label for marketing coffee and other produce and products produced as part of the illicit crop substitution process. In 2017 the first Wuasikamas store was opened in La Candelaria in Bogotá with the purpose of selling premium Nariño coffee from Aponte, handicrafts and panela. According to their website, 'one of the main objectives of this label is to provide a healthy life for the community and to support the productive infrastructures of Aponte, which after overcoming the armed conflict, also had to overcome losses due to the activation of a geological fault in 2015'.[23] While the built environment has since become completely destroyed by this phenomenon, reports by the UNDP indicate that the ecosystem rehabilitation process is well underway, with native flora and fauna 'in the process of being restored,'[24] including increasing populations of the iconic tapirs and spectacled bears.

2.3 Calamity on the horizon

The story of Aponte has for centuries been one of survival and perseverance in the face of conditions of permanent disaster and previous exposures to landslide disasters, leveling parts of the town. It was first in the spring of 2015 that the governor of the cabildo at the time sounded the alert after three houses had been affected by multiple, 'mysterious' cracks across both floor and walls. It was not long before all three families had to be evacuated due to the risk of structural collapse as the result of the ever-widening cracks that appeared in greater number every other day. What appeared to just be a curious, isolated phenomenon, however, turned out to be part of a wider and seemingly perpetual destruction process. At this point in time, no one could have predicted that the cracks would mean the beginning of the end for Aponte as it had been until that time.

By January 2016, a total of 170 houses were affected by cracks of varying severity and onset development; of these, ten had collapsed and three had to be demolished due to the risk of collapse.[25] Cracks as thin as a strand of hair could be as thick as a pencil only weeks later, which had at the time forced the evacuation of at least 45 households. Key institutions, such as the local school, a nursery and the local radio station, had also sustained heavy and gradually worsening damage. Infrastructures such as the sewage system, electric power supply and water supply were increasingly at risk and interrupted. According to inhabitants, one could hear cracking and creaking at night as the soil slowly shifted under the floors of their homes. As some locals put it, the town was being eaten away at by the soil, or was sinking into it, or sliding downhill.

Nobody knew exactly what was going on at the time. A commission of students from the Pedagogical and Technical University of Colombia had carried out a survey, and it seemed like the hazard had something to do with a geological fault or an instability in the land connected to heavy rainfall; a suspected active mass movement. Fieldwork and inspections carried out days later by representatives from the Gobernación de Nariño and the Mayor's Office in El Tablón had noted that the solution to the problem had not been decided on at the time, but that faults with a depth of more than four meters had been identified in the surrounding area, effectively dividing the community in two.

It was increasingly becoming evident that much of Aponte was about to become uninhabitable, and that if interventions were delayed further it was at risk of suffering the fate of Gramalote. The town of Gramalote was constructed on a geological fault line in the 1850s in a particularly hydrologically active site. During the extreme rainfalls associated with La Niña which devastated Colombia in 2010–11, a smaller earthquake coinciding with the heavy rains caused the soil to liquefy, effectively generating a slow-onset landslide that buried the town, killing few, but displacing nearly all of its 7,000 inhabitants.[26] While the Gramalote disaster is not remembered for its deadliness, the resettlement and community recovery process has been slow and ineffective and has raised doubts whether meaningful community recovery will be achieved by its displaced population.[27]

As it became increasingly clear that a complete resettlement would have to be undertaken also in the case of Aponte, involving the construction of an entirely new urban centre at an alternative site, life in Aponte became increasingly shaped by the sentiment of impending destruction and uprooting, not made better by the lack of political commitment for the relocation process. Even at the earliest stages of the resettlement planning process, stakeholders and response agencies highlighted the challenge of doing so in a way that respects the uses and customs of the Inga.

Thus, considering the characteristics of the geological phenomenon, Aponte had received a death sentence, which could evolve in at least two ways; either slowly and gradually, as had been the case until now, or it could at any time evolve into a rapid-onset hazard in the form of a massive landslide, potentially claiming many lives. In the words of one news report:

> Considering the characteristics of the mass movement, it may evolve into a flow of material with more destructive properties. In practical terms, there can be a collapse [of the fault] of enormous and devastating proportions. That

is to say, in addition to the serious structural damage of the houses with the associated risk of collapse, there is the additional threat of great earth mobilisations which could result in a major catastrophe. Hundreds of human lives are definitely in danger, and not only that, a whole project of human and social [value] which is definitely irreplaceable. Given this outlook and the behaviour of the landslide that is apparently very slow in time but fast in the conception of already inflicted damage (and that to come), the dimensions of the problem and the number of homes already affected, the institutions in charge as well as the community itself raised the question of resettlement to an alternative site.[28]

With this snapshot as a backdrop, the present book explores affective aspects of this outlook within and beyond the case context. Given that the original site where Aponte is located both is and continues to be ever more hazardous to live in, there is only minor outright opposition to resettlement. Still, from the point of view of how the process had evolved in mid-2016, the prospect of knowing with certainty that life in Aponte is threatened was characterised by a range of different sentiments, anxieties and worries both for the near- and long-term survival of the community and the Inga culture. These sentiments and the resulting questions they raised, which are to be discussed in this book, have wide-ranging implications not only for how we think about resettlement and community recovery, but also for how we think about disaster as a (socio-cultural) phenomenon.

Notes

1 My dissertation work centered on conceptualising the concept of disaster preparedness concept with a particular focus on uncertainty and ambiguity. El Niño became a phenomenon of interest for me after an unrelated trip to Colombia a few months prior to the fieldwork period as it stood out as a good case for preparedness for elusive hazards.

2 In short terms, one of the main thesis findings was that while the El Niño warm event of 2015–16 was slowly accumulating in the Pacific Ocean, few mitigating steps were taken on the ground owing to the uncertainty of impacts. Knowledge of the existence of a regional-level climatic phenomenon does not equal having time-specific and local knowledge about concrete hazard characteristics, making preparedness difficult. However, it is not impossible (for the full thesis, see: Staupe-Delgado, 2018).

3 This map was first used in Staupe-Delgado (2020) with permissions of reproduction in this volume given by Taylor & Francis Group..

4 An earthquake hazard struck the city of Popayán in 1983, killing over 260 people (see also: Gueri and Alzate, 1984). The earthquake occurred during the daytime, limiting its death toll. In the aftermath of the disaster legislation were passed requiring further improvements in seismic risk analysis and building codes, which have since been expanded on after more recent disasters.

5 The 2017 Mocoa landslide disaster killed over 300 people and left over 200 missing. It is considered one of the deadliest disasters in Colombia attributable to natural hazards in recent times. The reconstruction and recovery effort has been slow and controversial. At the time of writing many of the victims find themselves not having recovered fully with many families living in more vulnerable conditions than previously (one recent study of this particular disaster is for example Kuipers, Desportes and Hordijk, 2019).

6 See: https://www.unodc.org/unodc/en/frontpage/2018/September/coca-crops-in-colombia-at-all-time-high--unodc-report-finds. html

7 Official travel advice of many countries, including the Norwegian government, suggested that the rural areas outside of Pasto were red zone at the time. Word of mouth in Pasto also suggested that these zones should be avoided. Arranging for transportation was generally complicated for this reason.

8 The Unidad Nacional de Gestión de Riesgo de Desastres (UNGRD) is the central DRR authority in Colombia, and is also present departmentally through the various departmental government offices in the country, referred to as a Gobernación. The UNGRD has previously described in United Nations Office for Disaster Risk Reduction (UNDRR) reports as being a highly innovative institution (see Global Assessment Report 2019 (UNDRR, 2018)).

9 Cindoy and Chindoy (2017).

10 Ibid.

11 Ibid., p. 116.

12 Ibid., p. 117.

13 Ibid.

14 Ibid.

15 Ibid.

16 See: http://www.wuasikamas.org

17 Garcia (2019).

18 Ibid., p. 2.

19 Ibid., pp. 2–3.

20 Cindoy and Chindoy (2017: 119).

21 Ibid., p. 120

22 See: https://www.equatorinitiative.org/2017/05/30/wuasikamas-el-modelo-del-pueblo-inga-en-aponte/

23 See: http://www.wuasikamas.org

24 UNDP (2015: 7).

25 My synthesis from a various radio and video interview sources, including RCN, for which no URLs are available at present.

26 Displacement Solutions (2015).

27 News story in Spanish by Caracol Radio on insecurity and living conditions in the Gramalote resettled community: https://caracol.com.co/emisora/2019/09/27/cucuta/1569597485_008470.html

28 News story in Spanish by Fabio Arévalo writing for Las 2 Orillas describing also some of the potential hazard and disaster risk characteristics of the disaster: https://www.las2orillas.co/resguardo-indigena-de-aponte-narino-esta-colapsando/

3 The phenomenon

The natural hazard and its characteristics – some reflections

Disaster researchers generally distinguish between hazards and disasters by emphasising that the frequently catastrophic impacts of hazards depend on much more than merely their physical characteristics. Hazards are often understood as physical and natural processes, like landslides, although they can also be caused by man-made processes of environmental degradation. However, their destructive potential is largely a result of societal developments, and their adverse ramifications are often of a social nature. In this chapter I will outline the hazard characteristics of the natural hazard that is affecting Aponte. I find it necessary to clarify that although the basis for this account draws on geology, it will not be written from the perspective of a geologist. Rather, the aim is to focus on describing the defining characteristics of the hazard in order to facilitate the debates that will follow in later chapters. My objective, therefore, is to conceptualise the hazard rather than to provide a full geological description, with a particular emphasis on its gradual manifestation as well as the slow manifestation of its disaster impacts. Additionally, what sets this chapter apart from the two chapters that follow, is that Chapters 3 and 4 concern the experience of the hazard and the resulting slow calamity. In turn, Chapter 4 aims to capture the local conceptualisations of the nature of the hazard and Chapter 5 describes its slowly emerging disastrous impacts, or the disaster side of the coin.

As mentioned previously, the expedition to Aponte was in many ways a detour to my own fieldwork on the 2015–16 El Niño event, and I had few preconceptions about the kind of phenomenon we would encounter there. Therefore, to provide a necessary backdrop to this chapter, I will briefly present my initial impressions before elaborating with accounts from official sources, predominantly based on reports and conversations with the Colombian Geological Survey, in addition to supplementary data and information from other authorities and professionals, including representatives from Unidad Nacional para la Gestión del Riesgo de Desastres (UNGRD) and the Gobernación.

Upon arriving in Aponte, located on the hills of a green and lush Andean valley landscape, we were greeted by a small group of representatives who took us on a tour to survey the extent of the advancement of the phenomenon's onset as well as the destruction it had caused. The first sight that met us was the numerous cracks in the ground all over the plateau at the base of where Aponte is located. Most of them spanned about five metres in length and 20–25 centimetres in breadth, and

DOI: 10.4324/9780429288135-4

the descriptions I had heard in Pasto about how the town was 'cracking up' and 'being devoured by the soil' suddenly made sense to me. We proceeded past over a dozen homes and buildings that were covered in cracks of varying size and spread. More dramatic fault lines were also visible, some of which had caused buildings to collapse, splitting homes sharply in two. The difference in altitude on each side of the split caused one side to cave in, while the other side of the building often remained intact.

The longest fault line, and the one used to informally assess the progression of the hazard onset, effectively cut through the entire reserve, and stood as a wall around the uphill backside of Aponte's Nuevo Horizonte neighbourhood. The fault, a result of the sinking of the hill on which Aponte is situated, had caused the difference in altitude between the upper and lower sections to be well over a meter at the time of the expedition in the spring of 2016. Slowly but surely, the initial fault line grew steadily in height, and the 'cracks' witnessed throughout the town grew perpetually wider and more numerous. Potential material, social and cultural collapse was thus recognised as a very real threat to the community as inhabitants understood that their continued life in the area was threatened.

3.1 The hazard

On June 19, 2015, a letter was sent from the Gobernación requesting the office of the Colombian Geological Survey to initiate inquiries at the request of the head of the indigenous reserve of Aponte due to 'evidence related to the beginning of a possible mass movement process'.[1] The inquiry was carried out with the aim of assessing the extent and potential consequences of a stress-cracking process, which had, according to reports at the time, caused a limited number of homes to suffer structural damage in the form of small, but growing *grietas* (cracks) in walls, floors and ceilings, which was rendering affected homes increasingly dangerous to live in.

During field observations carried out by the Colombian Geological Survey on the spot, an eight-meter-long fault was detected which was considered indicative of a previous mass movement in the area. The previous mass movement process was stipulated to have occurred sometime around the year 2009 and 2013 with significant damage to the built environment. The inquiry also noted significant land-use change in the area in the form of vast coverage of non-native plant species on the slopes, but does not draw conclusions on the degree to which this deforestation may be an explanatory factor. Informally, however, deforestation and land-use change is considered a major factor in having destabilised the soil so as to increase disaster risk. Trees and other plants with deep roots serve to keep the soil in place and prevent it from being easily displaced. Additionally, the roots may limit the severity of mass movements when the soil does displace. Because of this, clearing hilly areas for farmland can end up destabilising slopes. In addition, Aponte turned out to have been situated on soils particularly susceptible to mass movements. Although the hazard affecting Aponte is also caused by rainfall and sedimentation, this process could have been significantly less severe had more deep rooted flora remained on the surrounding hills to keep them stable.

As part of the survey, the extent of the cracks, fault lines and total failure of infrastructure were observed to be worse than initial reports suggested, with clear indications that the onset was not yet advanced and that things could well degenerate catastrophically in the coming months or years (the onset dynamic was not easy to time, after all). The inhabitants of more recently affected homes told representatives from the Colombian Geological Survey that this process, first manifesting as tiny fissures, began towards the end of 2014 but that it was only in the middle of 2015 that it had caused the inhabitants distress, having by then expanded from hairline thin to pencil-thick cracks observable in the walls and floors of nearby homes. The initial long fault at the Nuevo Horizonte district was at this time evident, but without significant vertical soil displacement. Reports from the initial inquiry in 2015 noted the existence of several meter-deep cracks in the ground, often manifesting as clusters of cracks in a crater-like fashion. The inquiry goes on to suggest that the process is likely attributable to a lack of lateral confinement, related to the slope upon which Aponte is situated, which in turns triggers the appearance of stresses in the form of cracking, faults and their resulting material devastation as it displaces inflexible buildings atop it. In effect the hillside was sliding downhill in slow motion, and fears centred particularly on that it might speed up or collapse altogether.

In February 2016, a new emergency survey was made of the affected area with a number of representatives from relevant municipal and departmental agencies, at the request of the head of the Inga. During this second inquiry the initial fault line in the Nuevo Horizonte district was inspected once again, and it was noted that 'there is a considerable increase both in vertical and horizontal displacement', additionally stating, 'it is evident that the material on which the Aponte Indigenous Reserve is situated corresponds to colluvial deposits most likely due to material that have fell from higher up parts of the area'.[2] The survey concluded that there is evidence of a retrogressive process at play in the direction of the higher zones of the reserve—a conclusion drawn from comparing 2015 and 2016 observations. Hence, the implication is that 'the movement is active',[3] in the sense that new parallel and transverse faults appear and expand from the initial fault at Nuevo Horizonte. The phenomenon thus remains in motion, slowly worsening over time, and perpetually so. It thus became clear that parts of Aponte would not remain inhabitable for long, and that affected households would have to be relocated to an alternative site (at the time undetermined), preferably nearby.

3.2 Geomorphology and descriptions of the mass movement

According to the Colombian Geological Survey, the geomorphology of the Aponte urban area is characterised by an approx. 1,800-meter-long elongated cone at 2,400 meters altitude, with staggered soil deformations and undulations registered topographically in the highest sector. This landscape is dominated by steep concave and straight slopes of about 250 meters. These landscapes 'originate as hillside deposits, arranged at the base of the same, whose material is transported by gravity'[4] and loosened by interchanging dry and wetter periods. Hence, it can be inferred that the shape of the landscape is another contributing factor to the emergence of the hazard phenomenon, in combination with rainfall and runoff.

In the initial survey, the landscape or hillside on which Aponte is located was described as consisting of three well-marked steps or terraces, each about 10–20 meters high. The survey authors thus suggest that it may be 'inferred that this terracing marks the initial part of an older process; these relict characteristics are indicative of a deposit that has been affected in the past by mass movements'. It is likely, therefore, that similar mass movement processes occurred previously in the Aponte urban area before its current settlements were constructed.

The soil composition in the area consists of slimy-textured, brown, yellowish and red clay types and rich red soils of moderate firmness and consistency. A brief examination of the ground suggested that the soils go down quite deep before solid bedrock is reached. According to the inquiry carried out before my visit, the deposits which make up these hillsides are particularly susceptible to landslides, also indicated by the scars of previous historical landslides in the area. Experts note that such slopes, despite having been inactive for centuries, can be reactivated if the stability conditions change. As previously mentioned, the combined effects of land-use change, seismic activity, and changes in rainfall patterns are likely to have played a part, although the exact dynamic remains unclear and will be examined in greater detail in the following sub-section.

According to the Colombian Geological Survey, the current mass movement process (by February 2016) 'has a crown-to-tip difference of 30 meters, with a horizontal length of approximately 145 meters, with a pre-slope of 43° and a direction of movement of 40°'.[5] In addition, the agency estimates that the exposed zone constitutes an area of about 48,139 m^2 and that 'the thickness of the displaced mallet, taking into account the fault mechanism, is greater than 10 meters deep'.[6] It is feared that the mass movement process, whether it escalates or remains slow–onset, could displace a volume of sediment of over 480,000 m^3.

According to the inquiry, the current slow-onset mass movement process may be characterised by rotational slips, and described as that of an incipient generation, as it is of a retrogressive and advanced distribution and shows signs of widening in a leftward direction. It was also concluded that 'given the characteristics of the materials involved in the process, the humidity of materials, the morphology of the slope and their angle', the present slow-onset hazard process has the potential of changing its manifestation pattern into a rapid-onset type landslide—presumably upon reaching some sort of tipping point or critical weakness. The prospect of the slow hazard process turning into an immediately destructive and highly deadly disaster triggered fears that the onset dynamic would be of an exponential nature, as opposed to the previously assumed linear onset pattern.

Moreover, in line with current thinking on hazards and disasters, the natural hazard process should not be seen as unexpected nor its impacts unavoidable in any deterministic sense. According to the initial survey, there is a 'considerable density of mass movements' in the area, due to the sloping and soil composition of the valley. Contributing factors raising the exposure to such hazards are the drainage, general soil moisture and stream flows which interact with colluvial deposits. The previously numerous landslides, of smaller and greater size, have mainly affected agricultural lands, thus not triggering material destruction per se.

3.3 Initial theories and perspectives on the nature, onset and cause of the phenomenon

At the time of the field expedition, a range of explanatory models existed for the cause and origin of the geological phenomenon affecting Aponte. Important insights from the study of hazards and disasters have demonstrated that such phenomena tend to have complex patterns of causation, and that outcomes are more often than not a result of a convergence of factors. Diverging realist explanations for the origins of the phenomenon may therefore best be approached from the perspective that each represents partial truth. In other words, no single perspective offers an entire explanation for the event, but, nevertheless, each contributes an important piece to an overall understanding of the disaster. This will be expanded upon in the following subsection.

In terms of climate, the region of Andean Nariño is characterised by two marked rainy periods, the first during the three first months of the year and the second in the late fall, with January and November being the wettest months. The dry season lasts from June until August. However, according to interviews carried out as part of the February 2016 inquiry, the month of November in 2015 and that of January in 2016 were far drier than normal, with 'nearly no rain' in the month of January.[7] Because this dry climatic anomaly affected both wet seasons in the region, with such significant departures from normal rainfall patterns, the inquiry suggested that the anomaly is likely attributable to the evolution of the 2015–16 El Niño warm event which peaked at the time. Some also suggested a relation to the 2011 La Niña cold event. Moreover, given that the soil mass movement phenomenon continued to progress during this time, it was concluded that precipitation may not be the main factor driving its onset, thus stating that the phenomenon presents as 'dry'. According to the hydro-meteorological models and predictions, the spring season would be expected to be the rainiest one even during El Niño warm events, and under such conditions the prevailing theory is that that high levels of precipitation saturate the soil, which in turn affects pore pressure, causing a loss of resistance in the soil and thus accelerates the onset and extent of the mass movement. However, given the absence of local and reliable hydro-meteorological data, and the oral statements indicating drought conditions at the time, the exact role of rainfall is difficult to establish with certainty. However, rainfall is believed to be among the most significant explanatory factors. Also, rainfall entering the existing fissures perpetuate the process by speeding up the mass movement process.

Apart from precipitation, another common explanatory factor in the activation of mass movements is seismic activity, including earthquakes and volcanic activity. Nariño, located near the Pacific coast, in the so-called Ring of Fire, is exposed to significant seismic risk from both volcanoes and earthquakes. According to the Colombian Geological Survey there are various seismogenic sources with the potential to trigger the onset of the type of mass movement process which is destroying Aponte. The seismic activity does not necessarily need to happen nearby in order to trigger such processes, so long as they are sufficiently deep and potent enough to unsettle the terrain from afar. According to the seismic monitoring system in Colombia, there are no records of major earthquakes in the Aponte area within the time frame of the mass movement process. However, the Colombian

Geological Survey inquiry does note that such hazards are more susceptible to influence by seismic activity after their onset is already in motion. This means that in theory it is possible that seismic activity has affected the onset speed and severity, but is unlikely to have initiated the hazard in the first place. Given what we know about the convergence of factors affecting both hazard behaviours and disaster outcomes, these insights fit well within prevailing understandings of disaster dynamics. It also means that, theoretically, any seismic episode of a certain magnitude and depth may trigger other, seemingly unrelated hazard episodes years in advance through gradual processes of soil destabilisation, which, in turn, trigger very slow processes of mass movement, such as the one affecting Aponte.

In sum, at the time of the investigation which provided the inspiration and material for this book, numerous partial and potential explanations for the nature of the natural hazard phenomenon existed. Deforestation and land-use change may well have been an initiating factor by leaving the soil unstable due to an absence of deep roots keeping the landscape intact. Climatic anomalies which may or may not have been related to the influence of the El Niño warm event of 2015–16 may then have exacerbated the instability and caused further stress on the slope, speeding up the rate of movement as a result of changes in rainfall patterns. The onset speed also seems to increase as more rainfall enters into existing fissures and cracks. Seismic activities may also have been a contributing force, although geologists seem to agree that its cause is not of seismic origin. Nearby active volcanoes like Doña Juana produce frequent low-intensity volcanic tectonic events whose influence on the onset speed and pattern remains uncertain.

In the study of natural hazards, it is common for models and assumptions of causation to be speculative at worst, and uncertain at best–although critics of rationalist viewpoints suggest otherwise. Meteorologists, for example, are well aware of this challenge in their predictive models where the existence of even minor disturbances can generate major inaccuracies through what they term the 'butterfly effect'. In light of this, there are interesting analytical implications of the prevailing narratives surrounding the nature of the mass movement that is affecting Aponte. It is perhaps less relevant, at least for the purposes of this book, the degree to which these explanations are correct, but rather that they depend on the interaction of a number of factors, ranging from land-use change, rainfall, El Niño and seismic activity—or even some degree of 'randomness'.

3.4 Conceptualising the hazard profile

There are several characteristics about the hazard phenomenon which renders this case interesting for the topic of this book. As a starting point it must be declared that while the following is inspired by the hazard descriptions presented by Colombian authorities, this conceptualisation is only loosely based on these reports. Rather, the following subsection can be understood as an exercise in drawing out and abstracting the most analytically interesting hazard characteristics from the perspective of the wider field of disaster research.

The Aponte reserve is located in a hilly Andean terrain characterised by rich deep soil and considerable landslide exposure. In the previous subsections, we have seen that scars from past landslides are plentiful in the area, and that deforestation has further decreased the stability of the slopes. Although landslides are frequently thought of as rapid-onset hazards, the particular landslide affecting Aponte was characterised by a much slower onset, with about four years between the time the phenomenon was identified until the town of Aponte had nearly all but collapsed.

In the simplest sense, a landslide, regardless of onset, may be described as the downward displacement or movement of mass, usually down a slope in a hilly area, and may take many forms and causes. Their destructive potential is well known, as they may affect settlements built directly on top of or below the mass that is being displaced, as it moves quickly, and often unexpectedly, downhill, In the case of quick clay, lahars and more liquid types of landslides, their distance of travel can be considerable, meaning that settlements may well be exposed to landslide risk without being located immediately on or directly beneath a slope.

Aponte is located atop a plateau-like clearing near but not at the bottom of a relatively gently sloped valley. The previous section described the hazard phenomenon manifesting as 'cracks', and being of a 'retrogressive' and 'rotational' nature. Unpacking these terms helps us understand the hazardous phenomenon better, while leaving the technical geological aspects out. In this book I distinguish the hazard from the disaster in terms of impacts, meaning that the adverse impacts will be covered in the following chapter.

In 2015, when the first signs of the hazard were already evident, these were described in local media as manifesting as smaller cracks in the ground, which later also affected homes and infrastructure. In many ways, the emergence of these 'cracks', became the local way in which the phenomenon was conceptualised and the way that its environmental alteration and material destruction manifested itself. These cracks, in turn, indicated that the plateau was failing, slowly sliding downwards into the valley, with new cracks appearing every so often as the plateau gradually slumped into irregular-shaped terraces. It is important to keep in mind, however, that the perception of this phenomenon becomes elusive due to its long onset, as changes from day to day and week to week are nearly imperceptible. Much like watching a tree grow, while the change is obvious over extended periods of time, the daily observer often fails to appreciate the incremental developments.

As we saw in the excerpts from the Colombian Geological Survey in the previous subsection, the slow-onset mass movement phenomenon affecting Aponte is of a retrogressive nature. The retrogressive manifestation pattern of the phenomenon, although by no means uncommon for landslides in general, is of conceptual interest for the remainder of this book as it is relevant both for social perception of the manifestation dynamic of the hazard (as well as its impacts), and also for developing a better understanding for the hazard profile as such.

In ordinary language the concept of 'retrogression' connotes a degradation, movement in a backward direction, or, in other words, a process or advancement that can either be headed in a backward direction (retrogress) or be propelled by forces stemming 'from behind'. In geological terms, the concept denotes, at least in reference to landslides, that the mass started destabilising uphill, as opposed to on, say, a ledge, causing mass to be displaced front first in a wave-like fashion. This causes the mass to be displaced from the back to the front—in other words, once the critical weakness upslope collapses, it displaces the mass downhill. The Colombian Geological Survey further noted that the phenomenon was of a 'rotational' nature, meaning that the landscape is collapsing downwards and outwards in a concave shape. The retrogressive and rotational dynamic of the hazard is then what produces the scarps, such as the one depicted in Figure 2, a very standard illustration of how landslides function. The phenomenon affecting Aponte is similar, although happening in slow motion (at least for now).

Interestingly, mass movement processes manifest at multiple temporalities. Even for rapid-onset landslides, where soil seemingly instantly becomes unstable and comes crashing downhill, its onset and aftermath seems relatively clear-cut. However, even in this example the debris of the released landslide, now apparently stable, will often continue to move, sink and collapse. Simultaneously, new upslope mass from the scars in the terrain, still unstable, may continue to break off and come down, either as a result of minor influences like wind and moisture, or, more dramatically, by forces like seismic activity, for example. External forces in combination with critical weaknesses and stresses in the landscape thus ensure that eventually gravity itself will cause a collapse. Therefore, in the long term it makes sense to think of unstable slopes as existing in perpetual motion as the forces of nature, including gravity, continue to expose instabilities and critical weaknesses in the landscape until it eventually stabilises. Accordingly, applying a geological temporal perspective not only allows us to conceptualise even seemingly instantaneous types of mass movements in general as longue durée processes, but also sensitises us to the continued exposure to disaster risk, as the initial landslide is unlikely to have caused all the unstable mass to be dislocated at once, often leaving us with a considerable

Figure 2 Standard illustration of retrogressive landslide[8].

remaining disaster risk. The precise moment that the residual risk manifests as a new hazard is largely dependent on the extent to which the initial movement released the critical weaknesses in the slope.

Even though the geological hazard affecting Aponte is a slow-onset, creeping type of process, it can be productively framed as a relatively ordinary landslide in the earliest stages of its onset (indeed, rapid-onset landslides affected the territory on numerous occasions). In Aponte, the slope is still in the process of breaking loose and destabilising, slowly but surely generating momentum. We can therefore assume that eventually, once the weaknesses upslope have become sufficiently critical, the process will transform into a rapid-onset one when parts of the plateau collapse completely and come crashing down into the valley. It should be kept in mind, however, that for the purposes of this book the disaster onset period is focused on precisely this slow early-onset period. It was during this time that the settlement in Aponte was eventually further ruined as a result of the continual appearance of new cracks and slumps slowly but surely kept razing the built environment to the ground. In other words, from a disaster perspective, the disaster had already begun—and ended, at least in this location. Nevertheless, from a social perspective it may continue to linger as a disastrous experience for the population living in resettled areas. This means that conceptually it does make sense to distinguish between the onset dynamics of the hazardous phenomenon and the onset dynamics of its social consequences, particularly given the realisation that these two may oftentimes operate at significantly different timescales.

3.4.1 Challenges of perception and sense making

A number of characteristics of such creeping landslide hazards render them difficult to grasp by affected populations and untrained observers. Elusive and creeping environmental hazards may be approached through at least two types of epistemological lenses: one attempts to make sense of the world independently from culture and direct human observation and interpretation, as in the natural sciences; the second emphasises the way in which phenomena are lived, experienced day to day, recollected and rendered culturally significant.[9] Ambiguous environmental challenges such as these may in this way be seen as partially constructed discourses as they are interpreted, understood and re-negotiated as part of daily life and language.[10] Slow and gradual changes can often not be 'experienced directly by our senses, nor measured directly by our instruments'.[11] As with climate change and other creeping processes, then, such hazards tend to become 'a technical and abstract problem,'[12] while its consequences are suffered locally regardless of how the overall problem has been grasped and made sense of. A take-home lesson for the study of similarly elusive events is therefore that onset temporalities (gradual or fast, and the potential shift from one to the other), onset dynamics (temporary or permanent onset—stoppable or unstoppable onset, also connected to feasibility due to e.g. the cost of engineered solutions), and the attribution problem ('correctly' connecting the dots or distinguishing between causes and symptoms of the hazard—not to be confused with the causes and symptoms of disasters).

One of the problems of perception and attribution stems from the way in which the phenomenon manifests itself locally. Local residents report first having witnessed tiny cracks appearing in the ground and later in the floors and walls of their homes. Given their slow expansion both in size and number these changes are difficult to perceive on a day-to-day basis, at least initially.

A second problem of perception is the difficulty in knowing the degree to which the phenomenon is an event or a process. Each time 'the cracks' expanded and multiplied it could well have been the end of it—meaning that the onset would at that point in time discontinue. Essentially, at the initial stages of the onset it was by no means clear to local residents that the process was perpetual and unstoppable, and relative to other daily concerns minor cracking might not have been initially seen as a major hazard in its own right and certainly not a cause for declaring emergency.

A third problem of perception is the difficulty in realising that the individual sites of 'cracking' are mere manifestations or symptoms of the overall hazard process, which was the early onset landslide, and therefore do not constitute the main events in and by themselves. At the earliest stages of their appearance, this was not immediately clear and thus affected the way in which the hazard was first understood. In other words, the hazard was first conceptualised as being about the appearance of cracks which posed a problem. The ability to appreciate the overall disaster risk situation hinged in this way on the need to 'connect the dots' as a necessary condition for realising the graveness of the situation.

In conclusion, the appropriate appreciation of the seriousness of the situation is rendered elusive by a number of confounding factors. Chief among these are the very slow speed at which the 'cracks' evolved and multiplied, the impossibility of knowing that their evolution and multiplication was of a perpetual nature, and lastly that the individual cracking episodes were mere symptoms of the hazard, rather than being the hazard itself.

Notes

1 Servicio Geológico Colombiano (2016: 13).
2 Ibid., p. 14.
3 Ibid., p. 14.
4 Ibid., p. 36.
5 Ibid., p. 38.
6 Ibid.
7 Ibid.
8 I wish to thank Dominic Ochotorena for giving me this illustration.
9 Hulme, Dessai, Lorenzoni and Nelson (2009).
10 Kelman (2019b).
11 Hulme, Dessai, Lorenzoni and Nelson (2009).
12 DiFrancesco and Young (2011).

Part II
Experiencing slow calamity

4 Making sense of the hazard

Interpretations of the phenomenon

This chapter sets out to reflect on local understandings of the phenomenon that affected Aponte. Slow-onset disaster impacts often manifest along diffuse temporal and spatial dimensions, meaning that elusive forms of destruction and decay that occur gradually, but with no sign of dissipating, will necessarily produce different interpretations than instantly occurring phenomena.[2]

Since the inception of disaster research, the question of how societies and cultures understand and perceive hazardous processes has been considered a significant factor in analyses of structural disaster vulnerability. Central scholars in the field have pointed out the role played by ideologies, narratives and cultural frames in shaping disaster risk.[1] In the early work of Kenneth Hewitt, he remarks that 'the prevailing scientific view of [hazards] problems is a quite recent invention', by which he refers to a technical, 'partial and reconstructed view, carefully detached from almost all previous ideas of calamity, and reflecting the singular social context of its origins'.[2] Expert understanding of hazards are thus seen as woven into a detached vocabulary and formalised through technical bureaus, and as 'fully symptomatic of the social contexts in which it has arisen and that still form its main points of reference'.[3] The dominant view of hazards can then be seen as 'a construct reflecting the shaping hand of a contemporary social order', which, from a constructivist and cultural point of view, 'is itself a phenomenon requiring investigation as part of the so-called "social construction of knowledge"'.[4] Such a pursuit of insight on local conceptions of hazardous phenomena, as well as their oftentimes disastrous consequences, as discussed in Chapter 5, cannot be reduced to a mere curious interest in scientifically inaccurate risk perceptions to be corrected by risk communication:

> We are concerned with the way 'thought follows reality'. But the 'realities' here are not assumed universals of the empiricist's sense data and their psychological assimilation in acts of human perception and cognition. Rather we are looking at conditions that shape these pliable processes; the conditions that influence what facts we are likely to recognise and deem important; the acquired, accepted ways of interpreting them. These are matters of the social order.[5]

Thus, worldviews shape how hazardous processes, their manifestation and disaster potential are perceived, as well as prioritised, such as what forms of knowledge are

DOI: 10.4324/9780429288135-6

deemed acceptable and which untimely mitigative actions are pursued, if any. Local narratives about hazards can then be seen as 'a careful, pragmatic and disarming placement of the problem'.[6] According to expert perspectives, technical narratives may seem to be employed in order to 'maintain a sense of discontinuity or other-ness, which severs these problems from the rest of man–environment relationships and social life'.[7] The more 'foreign' these narratives seem, then, the greater the unproductive potential clash between worldviews may be, which most certainly will not facilitate mutual understanding and search for viable solutions. Such con-flicts have been described in terms of that 'the justification for the technocratic monologue is, of course, that the vast majority of [traditional] societies seem not to have the faintest idea ... Necessarily, therefore, the technocrat may presume to speak for these people, but can find little value in dialogue with them or learning from them'.[8] Disaster anthropologist Susan Hoffman also notes that perceived haz-ard characteristics may shape which hazards are emphasised by a particular culture and which are downplayed—it is 'culture that determines how a particular people calculate peril, experience catastrophes, and recover from them, or do not recover, or do not protect themselves'. As such, bridging local understandings of hazards and disaster risk is essential if meaningful solutions are to be found, and such work necessarily starts with identifying how hazards are locally interpreted and rendered culturally significant.[9]

As theorised in, for example, the work of Greg Bankoff, social constructions of hazards significantly shape views of causation in reference to deterministic or fatal-istic attitudes, thus they are of considerable relevance to the experience of the disaster process:

> The social construction of hazard in a society is not just of mere academic interest but should also be a matter of considerable moment for those engaged in disaster preparedness, management and relief. All too often, insufficient rec-ognition is accorded to the manner in which people's actions before, during and after a disaster are influenced by their cultural interpretation of [hazards], whether the event is perceived to be an avoidable climatic or seismic extreme or just a form of retribution meted out for a community's transgressions ... Religion remains a potent force to consider when dealing with popular expla-nations of the sudden and unforeseen. Nor are national governments com-pletely above such stratagems themselves, attributing their inaction and lack of political will to address pressing social and economic problems to the excuse of the almost godlike forces of an ungovernable Nature[10]

Indeed, for over five decades scholarship on hazards and disasters have attempted to bring to the fore that labelling the disastrous consequences of unchecked hazards as 'natural'—as in the term 'natural disaster'—, is part of such narratives, which at best constitute an unintentional misnomer and at worst a deliberate attempt to keep narratives agency-free,[11] or attribute the impacts to Nature. Other narratives that have been observed include those emphasising a 'sense of retribution', or 'the forces of Nature taking revenge for some imbalance or lack of harmony between human activity and the physical environment'.[12] Still other conceptions portray Nature as 'capricious', referring to the way in which it may be 'exhibiting uncontrollable characteristics that while not necessarily feminine, are often popularly associated with female or childlike behavior'.[13] Perhaps one of the best examples of a natural phenomenon that has been culturally mediated in such a fashion are El Niño warm events, referring to the Jesus child due to its observed tendency to peak around Christmas time.[14] From a functionalist perspective, the tendency to project human traits or frames of reference onto hazard phenomena may be considered an:

> [I]mportant means of maintaining cultural resilience in a society that experiences frequent disasters caused by natural hazards. It is a form of resilience because it represents an attempt by people to come to terms and deal with such phenomena by reducing 'the awesome and incomprehensible to something prosaic and simplistic' and so permits its incorporation within the structure of people's everyday cultural construction of reality[15]

Socio-cultural perceptions and constructs of hazard dynamics can in this way be approached as coping practices aiming to normalise such events. Whether these practices have emerged as ways of living with largely unmitigated disaster risks or as ways of cushioning their impacts is likely to depend on the context and phenomenon in question, including its particular manifestation dynamics, as well as underlying vulnerabilities. Either way, worldviews play a major role in establishing the kinds of normative systems, beliefs and practices that guide the direction of disaster risk—whether positive or negative in outcome.[16] Put differently, worldviews can manifest in effective or less effective coping practices.

Nonetheless, the degree to which local knowledge ought to be considered inherently valuable or 'in contrast to the notion of a body of "universal" and Western-scientific thought' remains 'a matter of considerable debate'.[17] More recent research on the matter, however, has leaned towards a pluralist approach, combining worldviews to solve problems scientifically in ways that are locally meaningful and desirable.[18] Still, a rich tradition of research on perceptions suggests that worldviews may also be seen as a considerable source of vulnerability, and prevailing narratives as shaped by power struggles or domination:

> The manner in which disasters caused by natural or human-induced hazards are conceptualized in the literature has matured considerably since the technocratic and physical explanations that were so ably criticized by Kenneth Hewitt and others in the 1980s. In particular, the analysis of a community's exposure to risk and their ability to deal with it has revolutionised the understanding of how and why such events occur. However, there is still a tendency to underestimate the extent to which disasters are also perceptual phenomena, occurrences that take place and shape in people's minds. The focus on people's physical, social, economic and political vulnerabilities and their comparable capacities or coping practices obscures just how much these are likewise cerebral events that influence behavior. Much of a people's resilience to withstand hazard lies in the intangible qualities generated by shared cultural attitudes and community spirit.[19]

Perceptions may also exacerbate feelings of hopelessness and fatalism and serve to undermine local coping capacities. Regardless, however, the role of perceptual conceptions of the ways in which hazards occur will materialise as concrete mitigative actions or cases of inaction. Hence, pre- and post-disaster strategies will be more likely to work if they are in collaboration with local frameworks and capacities, 'capitalising on them where they are strengths and attempting to moderate them where they may be obstacles'.[20] In this way, developing a better theoretical understanding for socio-cultural constructions of hazard manifestation dynamics will potentially be relevant not only for disaster theory in general, but also for concrete disaster risk reduction measures to be pursued.

Focusing particularly on the need for enhancing our theoretical understanding of such dynamics, disaster anthropologist Anthony Oliver-Smith argues for increased theoretical focus on the way in which hazards and disasters 'come into existence in both the material and the social worlds, and, perhaps, in some hybrid space between them'.[21] Hazards and their consequences are felt at the junction of environment and society, often vividly demonstrating 'the mutuality of each in the constitution of the other'.[22] The meanings attributed to hazards and their consequences, on the other hand, are shaped not only by cultural and social factors, but also by the concrete hazard manifestation dynamics, including onset speed, scale and severity:

> The array of disaster impacts from natural and technological hazards, ranging from rapid destruction and death from earthquakes to effects that go unperceived or unexperienced in the physical sense, often for many years, as in the case of toxic exposures, also encompasses wide variation. The variability of physical manifestations alone challenges our capacity to encompass the array of phenomena that generate and occur in disasters … Disasters are also both socially constructed and experienced differently by different groups and individuals, generating multiple interpretations of an event/process. A single disaster can fragment into different and conflicting sets of circumstances and interpretations according to the experience and identity of those affected.[23]

In this light it makes more sense to study disasters and the hazards that triggered them, in combination with underlying societal vulnerabilities, in the context of normalcy rather than as an exceptional event separate from pre-disaster (or pre-hazard manifestation) conditions. This agenda, still largely unrealised in the field of disaster research, traces its roots back to earlier scholarship, such as that of Kenneth Hewitt, and others before him. They argue that by shifting the analytical lens away from the disaster 'event' itself, in favor of a scholarly preoccupation with the social, cultural, political and environmental conditions—conditions that ultimately 'prefigure' or unleash the disastrous potential of hazards.[24] This shift towards disaster as a process rather than an event has, to some extent, shifted the scholarly assumptions concerning causation from the view that hazards oftentimes lead to disasters, towards a perspective in which the hazard is merely seen as a phenomenon that realises the latent disaster potential of a community. In other words, disasters are increasingly considered a result of long-standing processes of unchecked disaster risk creation and lack of mitigative measures, producing preventable disastrous outcomes in the face of even relatively low magnitude hazard phenomena.[25]

As we saw in Chapter 1, this has given rise to theoretical insights reached already in the 1980s that 'all disasters are slow onset'.[26] However, as is argued here, this perspective may not reflect local perceptions. While having had a profound impact on disaster scholarship in general and the vulnerability paradigm in particular, this and the following chapter set out to demonstrate that slowly emerging hazards produce disasters that are experienced—indeed lived—very differently than more instantaneous type of hazard manifestations.

For this reason, the focus of this chapter is on local constructions and narratives surrounding the nature, origin and causes of the hazardous phenomenon in Aponte. The next chapter is structured similarly, but with a focus on its material impacts—the 'disaster' side of the coin, while the rest of the book focuses on various cultural impacts.

4.1 The nature of the hazard

Hazards do unfortunately continue to produce disasters—according to many estimates at an increasing rather than a declining rate.[27] While disaster researchers have for the past four decades strived to shift attention away from natural hazards as the main explanatory variable in favour of the underlying conditions of susceptibility that render such events and processes disastrous, the specific characteristics of hazards, as well as how the hazard is understood locally, is still not insignificant for the study of disasters. Thus, writing hazards out of disaster theory in favour of an exclusive focus on societal conditions may have been premature, as hazard manifestation dynamics do shape disaster impact patterns to a significant extent. As the title of this book also suggests, gradually manifesting calamities are lived differently in the sense that destruction is temporally dispersed over a period of years rather than in the space of minutes or hours. While in both instances vulnerability is indeed a major part of the causal explanation, also in explaining the very long path to recovery for some, hazard characteristics as well as the onset speed of its resulting material and cultural destruction not only shape socio-cultural constructions of how life is lived throughout this process, but also influences response and relief strategies, including political saliency.

After all, the specific ways in which different types of hazards (still, unfortunately) impact societies not only shapes the way in which vulnerabilities are converted into direct losses and casualties, but also shapes their distribution of affectation,[28] again influencing cultural frames. Changes in the ontological underpinnings of hazardous processes and their oftentimes disastrous consequences thus lay the foundations for popular understandings of disaster risk and localised preparedness and response structures.

4.1.1 Representations in the media

News of the Aponte disaster has been covered by a range of media, including newspapers, radio, TV and social media. One can argue that the framings favoured by news outlets are to some degree shaped by initial public discourse on site, but

El pueblo indígena que se 'tragó' la tierra

Falla geológica amenaza con
desaparecer a Aponte Falla geológica amenaza con
desaparecer poblado nariñense
Grieta pone en riesgo a 170 familias en Nariño

Figure 3 A collection of headlines describing the phenomenon.

that the local discourse is also reflexively shaped by framings in the media. I will therefore start the discussion of local understandings of the hazard phenomenon by elucidating the meaning of a select few examples from newspapers and TV news coverage (Figure 3).

1. *The indigenous town swallowed by the Earth*
    Various news sources as well as aspects of the public discourse described the phenomenon occurring in Aponte in terms of the town being 'swallowed' or eaten away at by the soil. In fact, in the written press the phenomenon has hardly ever been covered as a mass movement-type of hazard, as the narrative has revolved around how the Aponte area is cracking up, being torn apart, sinking into the ground or crumbling—processes collectively described in terms of eating or swallowing metaphors.

    These types of metaphorical framings can be understood as partially dominated by a hazard-paradigm understanding, although the prevailing narratives are more complex than they first appear. A news article in the newspaper *Semana* which describes the phenomenon in terms of Aponte being 'swallowed' employs a mix of ontological frameworks. Contrast, for example, the following two excerpts from the same story:

    > Ironically, Mother Nature, who protects that community so much, has now become her main threat. It is now several months since the land of the shelter begun cracking up.

    > The phenomenon, which is investigated by experts, reached the main village of the Aponte district where the indigenous people live.

    The first excerpt refers to the story of Aponte described in the second chapter of this book, where the community had, through its embrace of the philosophy of Wuasikamas, declared themselves guardians of the land and keepers of the territory. It is also clear that many of the narratives, such as this one, that are presented in mainstream news outlets in the region render the phenomenon mystical by describing the process mainly though metaphors and a reliance on describing symptoms rather can causes (e.g. cracks rather than mass movement). At first sight, it might also seem like news outlets are attempting to narrate their story in a way consistent with the Andean cosmovición, but instead seemingly end up contradicting it, as will be discussed towards the end of this chapter. Furthermore, presenting the natural phenomenon as an external threat in this way ignores local narratives of environmental degradation as a contributing factor, along with a multifaceted understanding of potential causality.

The second excerpt refers to the inquiries being undertaken by the Colombian Geological Survey and a team of researchers from a local university to uncover the cause and severity of the hazard. Also here the nature of the hazard, which had been elaborated on in a report by the Geological Survey prior to the publication of this story, is left out of the narrative in favour of more ambiguous terms that obscure the nature of the phenomenon. What the article does convey accurately, however, by referring to how the phenomenon 'reached the main village', is the gradual expansion of the cracks in all directions and its retrogressive nature. These examples serve to illustrate some of the diverging ways in which the hazard process has been described in printed media.

2. *Geological fault threatens to make Aponte disappear*
We also see that the hazard is often framed in terms of 'the fault'. Confounding the meaning of the term, some observers were of the impression that this implies the hazard to be connected to tectonic plates, but instead it refers to the faults or scarps and cracks created by the onset of the mass movement. As mentioned in the previous chapter, the rotational nature of the movement causes the plateau on top of which Aponte is located to sink downwards and outwards, creating differences in heights manifesting as faults or scarps. Stress around the base or crown of the landslide then manifest as smaller cracks, and these move backwards up the hill in a retrogressive fashion.

The ultimate consequence of the faults described in the news article is the threat of the impending disappearance of Aponte, or its near-complete destruction if you will. It was therefore clear already in the beginning of 2016, when this particular article was written, that the hazard process had the potential to render Aponte uninhabitable. Notwithstanding its near-total destruction at its original site, the question still remains whether Aponte will successfully be rebuilt elsewhere, and whether, or the degree to which, this place will or will not be legitimately accepted as Aponte by a sufficient number of its former inhabitants. A theme explored towards the end of this book is for this reason the geographic boundedness of community. In other words, if Aponte is successfully rebuilt elsewhere and its population resettled, has Aponte disappeared or has it survived?

3. *Crack puts at risk 170 families in Nariño*
The notion of cracks, cracking and faults are, as we can see, a common theme running across many of the representations in the news. In essence, this is part of the reason why the hazard has been difficult to perceive and why it has been challenging to realise its urgency locally. The cracks are generally dispersed, evolving slowly and seemingly rather unthreatening, at least initially. In this particular headline the crack referred to is likely the main scarp, or the initial fault line described in the previous chapter, often used to assess the advance of the hazard process.

Also from this headline we can read that the overall mass movement process, or the macro-phenomenon, if you will, is left out of the narrative in favor of its immediately observable manifestations. This gives rise to a severe understatement of how many families are exposed in this context, here estimated at only 170, most likely due to a failure of appropriately assessing the correct scope of the entire phenomenon. Hence, and in contrast to the previous headline, which states that Aponte may well disappear, this headline significantly downplays the severity of the situation by omitting that the entire area is likely to be rendered more or less uninhabitable within a few years of the text in question being published.

4.2 Unpacking local narratives

Local narratives surrounding the hazardous phenomenon were diverse and informed by different sets of worldviews depending on who was consulted, as is reflected in this sub-chapter. Below is an analysis of some examples of how the hazard is conceptualised locally from various perspectives, including the way it is referred to, how it advanced and how it was experienced at the moment it became noticeable.

4.2.1 *Falla geológica*

Of the narratives I heard prior to the field expedition to Aponte, by far the most common ones described the process in terms of feeding metaphors; many drew on descriptions such as the town being 'devoured' by a 'geological fault'; others emphasised how Aponte was being 'eaten away at' or 'swallowed' by the soil. A recurring theme is that the hazard is conceptualised in terms of a 'geological fault' (*falla geológica*). In TV interviews during the early stages of the manifestation, for example, community representatives frequently describe the situation in terms of a 'geological fault affecting' the community. The presence of the hazard was also often described as having been inflicted by Mother Nature, which is a description that, if read in context, suggests a view in which agency is attributed to Nature with an undertone indicating punishment for not having lived according to the principles of *Wuasikamas*.

During my initial research, I recall first feeling confused about what kind of phenomenon this could be, as the description 'geological fault' did not provide me with an adequate understanding. The news media similarly, as shown above, focused mainly on framing the problem as the emergence of a long fault cutting across the Andean town. However, as readers might recall from Chapter 3, the fault often referred to is just the 'main scrap' in the landslide process, but by no means constitutes the entire natural hazard phenomenon. The initial tendency for the news media, people in the departmental capital of Pasto, as well as local leaders to describe the predicament of Aponte in terms of affectation by a 'geological fault' may well have served to undermine efforts at establishing urgency, thus reducing the political salience of the early political response.

4.2.2 It moves

An important aspect of capturing local conceptions of the hazard process is the way in which its onset is perceived, particularly with regard to causal factors shaping its speed and intensity. Such perceptions can either be in line with geological measurements or in stark contrast to them, but still tell us something about the experience of living with hazard. Numerous people attributed the progression of the hazard phenomenon to various environmental or climatic circumstances, as in the following observations:

- *When it's windy is when it moves*
- *When it's sunny it's worse*
- *It moves at night*

The excerpts above illustrate the tendency to explain the progression of the hazard process in terms of weather or time of day (I want to underline that I am not pointing this out in order to compare it to geological or 'scientific' explanations, but to illustrate points about disaster experience). In this way, the natural hazard phenomenon is understood as demonstrating an unpredictable onset that varies causally based on a number of external factors, such as the ones above. This tendency is arguably part of an initial phase of socio-cultural construction of hazard dynamics and causational explanation models, wherein affected populations strive to come to terms with their predicament——a process that starts with a search for explanations. These kinds of modes of attribution are most likely based on concrete observations and experiences, rather than as imaginative efforts at intuitive explanation. Such initial assumptions regarding causation are therefore likely to be adjusted continually by affected populations during the onset process and should therefore not be assumed to be static.

For the purposes of this book it is analytically interesting to think about the tendency of affected populations to explicitly theorise the onset dynamics of the hazard process. This tendency forces us to consider how causal claims are part of early onset hazard experience, without delving into functionalist reflections on the socio-cultural significance of such constructions (e.g. that certain perceptions might provide psychosocial consolidation rendering life in a hazardous environment possible, that they constitute strategies that enhance resilience, or a romanticisation of indigenous worldviews). From the time up until when expert narratives have become prevalent, pending a proper scientific diagnosis of the problem—or that the problem has been 'rendered technical', as described elsewhere[29]—affected households are left to reflect on their own perceptions of the problem. These perceptions should not, at least during early stages, be understood as representing a coherent community understanding and particularly not as representative of a grander *cosmovición* view of the process. Instead, causal claims such as the above allow us to grasp fragments of the experience of individuals, households and neighborhoods as they try to arrive at plausible explanations for what is going on, although these over time are likely to weave into a larger and larger explanatory framework or 'quilt' as people co-construct their hazard perception through conversation. These initial attempts at reflecting on causal explanations, however, should not be seen as final, representative or generalisable as they continually evolve and are adjusted throughout the onset period as new information becomes available and new and better explanatory frameworks appear.

Endogenous interpretations originating at the early stages of the onset of a particular slow-onset hazard typically reflect historical association or ways of thinking and imagining. Attributing or attempting to associate the lived experience of seeing ever widening cracks and geological faults with weather conditions, time of day and other factors thus indicate, at least in the beginning of the onset, that the progression of the hazard was not perceived as predictably incremental but rather as sporadic and perpetual.

4.2.3 Tremors, creaking and shaking

The dread aspect of the slow-onset manifestation of the hazard was a significant source of anxiety among the Inga. Every so often, according to members of the community, the ground would tremble and creak, causing fear and distress. At the early stage of the onset it was not fully clear how its disaster impacts would manifest, but the hazard phenomenon itself caused people to live with high degrees of uncertainty, as illustrated in the comments below:

-   *The phenomenon is very strong at night with these tremors that might destroy our house – that we hear at night – that it shakes – it is very difficult*
-   *Last night we felt them – the tremors are quite strong*
-   *Some nights it trembles and others not*
-   *Some people don't pay attention to it; they go on although it creaks*

As described by John Tulloch and Deborah Lupton[30] in a book on their comprehensive study on the lived experience of risk, gradual onset processes may fade in and out of attention. Often requiring explicit acts of noticing—paying attention to incremental changes in material impacts (e.g. slow changes in the breadth of cracks in the facade of a house), noticing changes in smell or sensation (in the case of toxicity etc.), or variations of sound (including intensity, character, duration and frequency)—the lived experience of such creeping disasters are shaped by a dualism of continuity and disruption. As the quotes above suggest, when the hazard process punctuates the everyday by producing noticeable sensory experiences such as creaks, trembling and so on, the immediacy of the ongoing hazard process is brought into attention again, sometimes sparking intense feelings of urgency and dread. At other times, the process may again be considered as existing only in the background, gradually building up destructive energy over time, yet overshadowed by the routines and demands of everyday life which still has to be lived during the onset phase. People, after all, continue to engage in their livelihoods, strive to keep their children in school and attend to the sick and elderly throughout the process, at least until impacts no longer permit it.

4.3 Descriptions of the progression of the hazard

As the preceding sections attempt to make clear, we may speak of various temporalities concerning both hazards and disasters. In a typological sense, therefore, we may consider slow-onset hazards producing rapid-onset consequences (a slow landslide suddenly breaking off to produce a tsunami), slow-onset hazards producing slow-onset consequences (e.g. droughts and their effects), as well as rapid-onset hazards producing rapid-onset impacts (earthquakes for instance), and indeed also rapid-onset hazards producing slow-onset impacts (e.g. certain toxic or radiation incidents). A reasonable initial assumption is to assume that by developing a better understanding of the way in which early signs of hazard manifestations are experienced and made sense of locally helps us better understand how the subsequent disaster impacts are felt and coped with.

4.3.1 Earliest signs

Slow-onset hazards are often not experienced directly until they are already well advanced. Eclipsed by the lived experience of everyday life, the earliest signs are often overlooked for various reasons, at least for some time during their early onset. In Aponte, the hazard phenomenon first manifested as smaller cracks in the ground and smaller faults in and around the town. In this way slow-onset hazards are first felt mainly as a minor or even insignificant nuisance. When it later became clear that the faults were gradually growing and with new ones appearing, it became evident that this was not merely a top soil problem but also a geological hazard. Thus, as hazard onsets advance, their initial manifestation outside the realm of everyday experience transforms gradually into observable alterations of the environment. While these manifestations may be ignored for some time, or more realistically will pass in and out of perception and lived experience as the chores of everyday life distract from them, their presence eventually becomes omnipresent—hard to overlook and a defining feature of the experience. It is particularly when their adverse impacts are also felt that they eventually gain the potential to disrupt life as it was lived until that point, although thresholds for when a gradual phenomenon reaches emergency status can be difficult to determine and agree on.

4.3.2 Connecting the dots: from events to process

Elusive phenomena often manifest as seemingly distinct events. These are events that 'simmer' in the background, with changes which occur so slowly that perceiving their interconnectedness and their future potential severity is rarely possible for the untrained eye. Connecting the dots, or attributing the individual manifestations witnessed on the ground to the overall macro-process, is largely a process of conceptual abstraction and reliably attributing the parts to a greater whole. Akin to what in climate research is labelled 'parameterisation',[31] establishing relationships between local episodes of drought or floods to a greater climatic anomaly, such as El Niño, suggests that such measurement and experience is in part a product of meaning construction. Realising the scope of such processes, then, hinges on local populations making correct assumptions about 'symptoms' so as to perceive them as part of an overall 'illness', rather than as one-off events.

In Aponte, it was not clear from the initial stages of the onset process that the settlement would eventually face complete ruination. What started as faults at random sites around the reserve scarcely suggested the eventual end of continued existence in that place, at least from the perspective of local lived experience. One problem with observations is that when it eventually becomes evident that seemingly distinct events are part of a macro-phenomenon or process, the onset may already be quite advanced. However, in the case of the hazard affecting Aponte there were not many mitigative courses of action available, so having detected the severity of the phenomenon at an earlier stage would not have affected the ultimate outcome of displacement and resettlement.

4.3.3 Feelings of uncertainty about the hazard's progression

Knowing that a particular hazard is unfolding gives rise to a feeling of imminence and dread despite uncertainty about its exact onset dynamic and speed. From a phenomenological perspective, the manifestation dynamic of slow-onset hazards may be described in terms of a prolonged cliff-hanger experience. At least, in the absence of certainty about the continued development of a hazard process, affected populations may remain hopeful that each new development is the last and that things may gradually return to normal. Once it becomes clear that things will not, however, the lived experience changes from one characterised by living with the hope of recovery to one of dreading the inevitable end of life as it had previously been lived.

A point I wish to make here and in particular later in the more analytical chapters of the book is that, upon knowing, or realising that the future will bring only worsening conditions gives rise to a sense of 'living in anticipation' of an inevitably bleak ending. Receiving and comprehending the fact that life in a place may not continue due to imminent disaster—or indeed the gradual unfolding and intensification of impacts in the present—punctuates daily routines and hopes in a way that renders the creeping hazard a defining feature of lived experience. It also gives rise to the realisation that for the time being life has to continue to be temporarily lived *in situ* as livelihoods must be attended to and daily life must be maintained until alternatives are on the table. It makes sense, therefore, to say that even in lives shaped by anticipating impending some degree of normalcy also exists. This omnipresence of the process is equally disruptive as that of adversities that are suffered in more explosive ways, yet is lived differently due to its lack of acuteness. Feelings of hopelessness, despair, of being doomed to face inevitable adversities and uncertainties nevertheless generate intense feelings of distress, perhaps particularly as this state of affairs can persist for years until solutions offering new hope are available.

4.3.4 Noticing and perceiving

Seemingly unrelated and gradually manifesting phenomena, such as the one affecting Aponte, exist, at least initially, only in a technical sense through measurements and expert systems. In the busy schedule of everyday life noticing and perceiving gradual change is difficult and early detection of disaster precursors therefore hinge on noticing and perceiving in an active sense—striving to notice and striving to perceive gradual changes which are not immediately obvious unless repeatedly observed over time. Acts of noticing and efforts at actively perceiving are further complicated by cultural and mental models that normalise or misrepresent such changes, and when their full severity is finally realised the onset may well be relatively advanced.

This challenge connects to the philosophy of science underpinning all phenomena that the sciences study. In their essence, natural phenomena behave in set mechanical (although by no means linear) ways in the physical sense. On the other side of the equation we have our perception of natural phenomena, or the way in which they are socially constructed, interpreted and culturally rendered meaningful. From such a critical realist perspective, it thus makes sense to say that the challenge for scientists, practitioners and affected populations alike is to be able to approach and understand the natural phenomenon in a way that facilitates appropriate measures. Misgauging or misdiagnosing the underlying hazard process as well as the social context, one could well argue, is to risk pursuing inappropriate responses, which is a source of vulnerability in itself. Following this line of reasoning it is worth underlining that in order for appropriate actions to be taken the underlying hazard process does not need to be fully understood in the scientific sense—partial understanding is often enough. After all, sailors successfully navigated the seas for millennia based on the assumption that the Earth is at the center of the solar system. Therefore, in discussing acts of noticing and drawing appropriate inferences, affected populations do not necessarily require a full understanding of the technical aspects of the forces at work, which are insights that scientists often also lack when a strict knowledge criterion is applied. Instead, correctly inferring that events are frequently processes, and that seemingly individual observations may be part of the same macro-phenomenon and noticing the difference between causes and symptoms are often enough to significantly reduce disaster risk. From the perspective of indigenous knowledge, then, acts of noticing and the drawing of appropriate inferences do not need to be based on highly technical rationales for it to result in mitigative action; it suffices that at-risk populations are realistically aware and reflexive about their own disaster exposure. A grave injustice would then be to silence such endogenous interpretive processes and concerns, as we so dramatically witnessed in the context of the 2009 L'Aquila earthquake and its aftermath.[32]

4.4 Perspectives on the phenomenon and dynamics of causality

It is (quite obviously) normal for populations affected by slow-onset natural hazards to reflect on the continued sets of risks they may be exposed to as the hazard keeps advancing, the nature of the hazard in terms of both onset and scale, as well as engaging in critical conversations on the issue of causality. Local disasters often become defining events in the history of a place and are thus naturally a topic of conversation and reflection. In Aponte, it was clear that the population expended a lot of effort reflecting not only on their situation, but also on the slow-onset phenomenon they were experiencing. Such reflections centered on factors that were believed to shape the onset, its continued development, potentially mitigate measures as well as how their relationship to the natural environment may have shaped disaster risk.

One such perceived aspect had to do with the hour of the day. It had been observed that advances in the onset of the phenomenon were most marked during nighttime. Affected households reported having occasionally heard creaks as they tried to sleep or having perceived cracks being larger than the previous day when waking up in the morning. It is perhaps quite logical that perception would lean towards emphasising the time of day as central to the onset. After all, it is during nighttime that the risk of being trapped in the rubble of one's collapsing home is greatest. Either way, the fact that many affected families perceived the phenomenon to be more active during nighttime suggests a tendency for night and day to feature actively as part of the sense-making process. It may well also be that the presence of threat is more intensely experienced during nighttime due to the enhanced perception of risk of adverse consequences during sleep, thus causing people to notice the onset more, including creaks and tremors resulting from the creeping progression of the phenomenon.

Weather- and climate-related conditions are another aspect that is perceived to affect the onset dynamic of the phenomenon. Whilst some suggested the progression of the hazard process was connected to wind conditions, others wondered if it was not perhaps worse on sunny days. El Niño was also suggested as a contributing factor, along with its associated rainfall patterns. A small minority of informants also stressed the potential role of climate change in driving the hazard onset forward. While precipitation patterns clearly play a role in the hazard onset, it is not my aim here to reflect on the validation or debunking of the prevailing causal reflections. My goal here is rather to bring to light the active ways in which acts of noticing and attempts at making sense of the slow changes observed are actively engaged in and negotiated locally. Of particular interest perhaps is the way that people adjust and incrementally adapt prevailing theories surrounding the onset based on new observations, and upon realising that previous conclusions do not hold up to observations made thereafter. For example, at the time of the field expedition in 2016 the El Niño phenomenon was still active and much discussed in conventional media. This may have led to El Niño being hypothesised locally as a potential causal factor and featured in several discussions surrounding the nature of the phenomenon. However, when the onset continued after the passing of El Niño, it eventually became clear that while changing rainfall patterns during the El Niño episode may have had a triggering effect, the passing of El Niño did not impact the onset of the geological phenomenon, as observed through the continued widening of cracks in the ground and in the built environment. It is worth noting, therefore, that elusive environmental phenomena will tend to trigger reflections and intense discussions in affected communities concerning the nature of observed symptoms.

There were also epistemological disagreements connected to generational differences. Elders would often emphasise the lack of enthusiasm for ancient practices among the youth. At the same time, youth occasionally expressed a lack of interest in the cosmological vision. One household expressed that the consumption of hallucinogens is declining, and that people to a lesser extent express the explanations arrived at through spiritual logics. Some emphasised that the elders had attributed the presence of the phenomenon to cultural change, that they had failed to notice the deteriorating state of the natural environment of their territory—partially due to failures of noticing and perceiving—where hallucinogens is by elders seen as playing a major part. Youth, on the other hand, were framed as less interested in rituals. While hallucinogens were understood as an important path towards clarity in the context of the phenomenon, many also mix or reject their importance in favor of the more technical explanations communicated by local authorities. Common for them both is that they arrive at environmental degradation as an important driver of Aponte's exposure to the geological hazard process.

As mentioned elsewhere, spiritual practices in the region draw on a mix of Catholicism and regional variants of indigenous beliefs and religious practices. The Christian God is thus also a central part of religious and spiritual life. However, the calamity has greatly shaken faith. While some question how such a thing could be allowed to happen and what has been done to deserve such a thing, others saw the continued onset of the phenomenon as due in part to loss of faith in God and other religious practices. Past mistakes, such as extensive poppy cultivation and environmental exploitation, were seen as indicators that old ways had been lost. The natural hazard hence came to be seen as a wake-up call. External customs, based on a profit-maximising logic and imported from abroad, had corrupted local values and spiritualties and thus shaken connections not only to the divine but also to Mother Earth. Some thus saw repentance for past mistakes as central, but doubt existed over whether such a move would stop the onset of the hazard, or if the disaster was imminent as a form of punishment. Community representatives have thus argued that the outcome of the disaster recovery effort will be the next great test of the Inga.

While the onset has advanced slowly, TV and radio interviews carried out at later stages of the onset in 2018 and 2019, years after fieldwork, indicate that people experienced the temporality of the situation in uneven ways. Looking back, affected people are surprised at how fast the phenomenon has progressed. In hindsight, the seemingly gradual changes, which are hardly visible on a day-to-day basis, accumulate into images of vast material destruction that hardly seemed imaginable at the initial phase of the onset. The slow onset thus made it possible to bracket out the disaster from everyday experience, while at the same time having it seep in at least a few times a day as conversation topics shift towards the material and observable impacts. One defining feature of the hazard process may thus be that they enter in and out of consciousness; never truly threatening and yet never ignorable.

Notes

1 One example is the oft-cited Pressure and Release (PAR) model (Wisner et al., 2004), which arguably serves to illustrate visually how we commonly understand the abstract concept of vulnerability. The model provides a conceptual framework that views disasters not so much as a result of hazards as in terms of ideological, societal and practical circumstances. At the highest level of abstraction, disasters can be seen as rooted in prevailing systems of power, systems and culture. These dynamics are often challenging to unravel due to their ubiquitous and taken-for-granted nature. Examples include how we relate to means of production (capitalism), the global system as well as balances of power and highly abstract narratives about the world and human nature, as well as the nature of disasters (the view that disasters are acts of God or that disasters are attributable to nature may perhaps qualify as examples of such taken-for-granted ideologies). In many ways, these ideological and culturally embedded ontological systems are too broad for direct scholarly inquiry, and researchers themselves are arguably too imbedded in them as well for being able to fully untangle their effects in precise ways. At a more concrete and middle-range level of abstraction we find the way in which societies and their intuitions are organised, including disaster management systems. At this level we also find societal macro-forces such as demographic changes, urbanisation pressures, economic policies, social policies and other macro forces that are still relatively concrete in comparison to the more elusive ideological factors that ultimately steer these societal variables. At the micro-level, the model also theorises that these dynamic pressures in turn end up shaping how people live, the quality of mitigation and preparedness measures, the existence of warning systems, the nature of livelihoods and other hands-on variables, and that the sum of all these three tiers of explanation shape disaster outcomes and disaster experience. Several alternatives also exist. Some examples of related frameworks attempting to elaborate on the role of vulnerability in disaster risk creation include: Turner II et al. (2003) and Birkmann et al. (2013).
2 Hewitt (1983: 3).
3 Ibid., p. 4.
4 Ibid., p. 4.
5 Ibid., p. 8.
6 Ibid., p. 10.
7 Ibid., p. 10.
8 Ibid., p. 18.
9 Kelman, Mercer and Gaillard (2012).
10 Bankoff (2004b: 92).
11 Chmutina and von Meding (2019).
12 Bankoff (2004b: 95).
13 Ibid., p. 96.
14 Ibid.
15 Ibid., p. 97.
16 Wisner, Gaillard and Kelman (2012).
17 Ibid., p. 110.
18 Balay-As, Marlowe and Gaillard (2018); Mercer, Kelman, Taranis and Suchet-Pearson (2010).
19 Bankoff (2004b: 110).
20 Ibid.
21 Oliver-Smith (2002: 24).
22 Ibid.
23 Ibid., p. 25.
24 Hewitt (1983: 27).
25 Wisner, Gaillard and Kelman (2012).
26 Lewis (1988.4).

27 For example: https://www.munichre.com/en/risks/natural-disasters-losses-are-trending-upwards.html#-1624621007

28 Although hazards are not what ultimately cause disasters, their onset dynamics as well as the intensity of the physical forces released will, to a significant extent, shape outcomes. This is not the same as saying socio-political processes of disaster risk creation are not the main casual component in explaining why the hazard became a disaster in the first place. However, when disasters do occur, certain hazard-related attributes are still of relevance and is also recognised in casual models such as the PAR model. An at-risk population will ultimately suffer greater losses if the hazard is stronger than they will if the hazard is weaker. Similarly, the way in which the resulting havoc is wrecked and experienced will depend, among other things, on the onset speed of the havoc.

29 Li (2007: 7).

30 Tulloch and Lupton (2003).

31 Kelman (2019b).

32 For more on the L'Aquila disaster, see for example: Alexander (2014) and Imperiale and Vanclay (2019).

5 Living with a slow calamity

Disruption and continuity in the face of creeping destruction

We have already seen that the disaster was initially experienced as the slow emergence and incremental worsening of 'cracks' in buildings and roads. The natural hazard phenomenon, or the incremental sinking of the territory into the valley over which it is situated, renders the ground unstable as it is in constant movement, albeit at a slow speed. Shifting the focus to the societal impact of this state of affairs, the impact is mainly felt through the gradual collapse of the built environment. As this happens day by day, little by little, in front of the people who live there, life increasingly becomes consumed by an imminent sense that the place where they based their lives and identity is doomed. As some houses collapse while others remain due to the uneven spatial dynamic of the hazard onset, the disastrous consequences are also experienced differentially. However, the totalising nature of the phenomenon meant that those who witnessed the collapse of the home of their neighbours were not in a position to sigh in relief that they themselves remained unaffected; it was universally understood that no one was safe and that the differential impacts were mainly temporal in nature. Eventually, although at an uneven pace, it became clear that everyone would find themselves both at risk of injury or death in their slowly collapsing homes, while at the same time living with the certainty that potential displacement would follow unless a well-managed resettlement plan was agreed upon in time.

In an interview in the local news from mid-2017 a community representative vividly captures this sentiment.[1] The story presents a woman standing by the ruins of her former house, which had collapsed at an early stage of the onset. Overlooking the rubble she points in the direction of where the view from her window used to be. She recalls being able to look over the painted houses of Aponte as she was sat by the window, where she enjoyed her view over large parts of the town. As a community leader, she had appreciated being able to observe what moved in Aponte and appreciated the view from her house, which allowed her to observe on a daily basis the until-then continuous improvements all of Aponte had witnessed over the years. She recalls and misses neighbours coming over for a chat and to greet the family.

For two years she has lived with the uncertainty of how the process will turn out, witnessing how her home and those of others have gradually been destroyed. Clearly mourning the loss of her home she expresses 'feeling a sense of desolation' and melancholy upon seeing the ruins of her home and those of some of her

DOI: 10.4324/9780429288135-7

previous neighbours, and also feeling upset for those who have not yet lost their homes but eventually will. She recalls memories of her neighbours being a central part of her life, being there when she woke up and noting that now the community is scattered as a result of the ongoing disaster. She gazes over the half-destroyed towinn, underlining the painful feeling of seeing the damaged homes, some with people still living in them, and some families that have long since vanished. Her own house was new, less than two years old, and she recalls happy memories living in it together with her four children and husband until the soil 'ate it'. All that remains is a ruin comprising four walls and the foundation of the house, all of which are full of large cracks.

She points out that the recovery process was not going as anticipated, and that many initial surveys and investigations, including social and vulnerability analyses, remain (bearing in mind that the interview is from 2017). She highlights that the Inga have helped the government before by eradicating their lands of illicit crops and resisting guerrilla interference, and expect the government to take their needs seriously. She notes how the Inga do not feel the process has been proactive, given that many have been displaced and are still waiting for shelters and relocation plans to materialise. She goes on to reflect on how the community has remained resilient in the face of many adversities in the past—including guerrilla attacks, the presence of paramilitary groups, deaths, political neglect, and other consequences of a war that was never theirs—and will for this reason strive to remain resilient in the face of current adversities posed by this phenomenon of nature.

While anecdotal, the story above gives us an impression of the phenomenological sense of living in an environment characterised by anxiety, uncertainty and melancholia, which, I argue, are defining traits of the lived experience of creeping disasters. We see that there is a sense of uneven patterns of impacts by the hazard onset; people go through disaster phases unevenly,[2] some having already been displaced while others live in anticipation of being displaced in the coming months. The Inga have faced adversities with resilience in the past, but nevertheless resilience in the face of an unstoppable natural phenomenon seems difficult to conceptualise. This gives rise to the question of whether resilience makes sense in such contexts and if or how resettlement can be understood as resilience. After all, a community cannot simply be moved from one place to another, the survival of the Inga hinges on being able to recreate similar social dynamics at an alternative location——a process which, in turn, depends on being able to maintain the relationships between people despite periods of displacement and hardship faced in the meantime. We will thus shift our attention to an elaboration of the everyday experience of living with the slowly emerging disaster impacts with the purpose of better narrating this dreaded phenomenological sense of living in anticipation of impending adversities.

5.1 Disaster impacts

As we saw in Chapter 3, the slow-onset hazard affecting Aponte would ultimately lead to a complete destruction of the built environment. In many ways one could argue that the resulting experience of loss and calamity is mainly attributable to exposure, or the fact that the town of Aponte was located at a site with such a hazard potential. However, taking a phenomenological view of disaster means going beyond looking at attributions and causal explanations by instead focusing the analysis on how the disaster is 'lived', as I have done in this book. On several occasions throughout this book I have thus used the term 'calamity' to describe the Aponte disaster, as this most closely resembles lived experience. After all, the Inga experience the impacts of the disaster as totalising, and as a threat to not only present well-bring but also the survival of their community. Put differently, the disaster has impacted the community in different ways.

The material aspect of the disaster is its most immediately visible aspect. The experience of either having lost one's home or living in a home that is slowly but steadily crumbling is clearly a precarious experience. On the one hand, the families who had already lost their homes stress the melancholic feeling of no longer being able to live where they used to live, whilst families whose houses were damaged and are now undergoing a slow collapse feared both for their future housing situation and, in some cases, for their lives. While material loss is mainly thought of as losing one's home, the emotional effects also include the anxiety surrounding losing one's sense of place, conception of home, networks, identity and current living standards. The uncertainties surrounding the shelter situation and resettlement programme exacerbated these emotional effects as people did not feel confident that the process would be well managed. A prevailing fear was that the quality of life and living standards would be significantly lower.

One aspect of the disaster that is not as immediately visible, but which is perhaps even more pressing, is its potential cultural impact. The disaster, even at its early onset, gradually displaces people who either go on to live in (reportedly inadequate and unsanitary) temporary shelters or who move elsewhere to live with relatives or start new lives. As a result of displacement, disasters have for some time been known to be a major factor in rural-urban migration as well as other forms of migration. Whilst the experience of displacement was traumatic for those affected, the impending displacement of those still remaining also dominates emotional life. Worries that the culture will not persist or will shift too strongly in an undesirable direction are strong and well founded. With some people having left to find opportunities elsewhere, it is unclear how many households need to remain for the community to persist as it is remembered by those who lived there.

Livelihoods are also adversely impacted by the disaster, with many worried they may end up poorer. Life savings and years of investment have gone into houses that are now either destroyed or crumbling. In the absence of concise talks about the resettlement plans and options, many affected households were convinced that reconstructed houses would be of a lower standard or that the resettlement process would be suboptimal in some other way. While Aponte is not a rich community, its relatively egalitarian structure, based on strong communitarian values, has led the Inga to enjoy a relatively comfortable life, albeit on considered poor (in the material sense) from the perspective of outside experts. Also, although many residents will be able to continue growing coffee despite the disaster, as the coffee fields will still be mostly usable, others are worried about idleness. For people who will have to live in shelters or who were small business owners, it is unclear if losses will be sufficiently compensated for them to be able to make yet another investment to keep their business alive after a resettlement process.

Affective and emotional aspects of the disaster are also strong and represent a potential source of trauma. The experience of slow calamity is perhaps most intensely lived as a kind of dread, in part based on an anticipated future feeling of melancholia. In striving to live and savour the community while it is still relatively intact, people indirectly end up letting an expected (negative) future dominate present affective states. The emotional toll of worrying about impending displacement, risks associated with gradual material destruction, potentially worsening socio-economic conditions in the future, as well as fear of cultural deterioration, was a major source of stress for many. The disaster in this way impacts life and lived experience in many ways, ranging from concrete to abstract forms of suffering, as outlined above. We will now move on to reflect on other aspects of the lived experience of the disaster.

5.2 Disaster impacts and lived experience: recurring themes

Narrating and capturing the exact ways in which people were affected by the Aponte disaster is a difficult task as it would depend on the ever-changing circumstances occurring throughout its incremental onset. Many households already saw their houses collapse or rendered too risky to live in by early 2016, whilst the homes of others were intact over a year later. A chronological type of account is thus confounded both by the onset and the spatio-temporal distribution of damages and their resulting social hardship. In this text I will therefore mainly focus on the lived experience of impacts without being particularly mindful about the order of occurrence of things. Instead, the subsection will present information as a mosaic of fragments of experiences, thematically organised, with the aim of painting a fuller picture of the phenomenology of slow calamity.

5.2.1 Surprise and non-linearity

In late 2015, affected communities reported having first experienced the slow-onset geological hazard (the natural phenomenon driving forward the slow calamity in the form of social, cultural and livelihood losses) in the form of faults appearing in the ground and cracks appearing in the foundations of homes and streets. A recurring theme is the tendency of affected households to initially believe the expansion of this phenomenon to be relatively linear, while maintaining an awareness of the fact that if homes were to eventually become structurally vulnerable as a result, their lives might be at risk.

One woman recalls having first paid little attention to the small cracks appearing first in the floor of her house, especially in and around corners. She reports feelings of surprise and disbelief that these smaller cracks would later turn out to be initial signs of a calamitous event for her community. The time it took for the seemingly innocent cracks to creep up the walls, cut across the roof and for the house to become structurally compromised was shorter than anyone could have anticipated.

Another family reports having struggled to make sense of the uncertainty they experienced, both as part of the onset dynamics and in planning for socio-economic contingencies. Living in uncertainty about what may happen next at any given time was a big source of distress. It was stated frequently that not knowing if everything would collapse increasingly quickly, or whether things would continue to progress gradually, was also a major source of anxiety. People had come to realise that the onset was not linear nor predictable, giving rise to fear about the realism of both official and private contingency planning efforts.

A common thread appears to be the lack of what we may label 'acuteness', which is arguably a defining feature of slow-onset types of disasters. Impacts emerge and cause severe consequences, but these are not delivered all at once. They are instead suffered gradually, with perhaps their worst psychosocial and cultural impact being their intense ability to produce dread and anxiety connected to uncertainty surrounding their continued, future onset trajectory. As in Aponte, we see that affected households describe how the difficulty of experiencing any kind of continuity or predictability greatly reduces their lived sense of security; the intense insecurity experienced by realising that the future may present a continued set of adversities that cannot easily be controlled or predicted by individuals. While earthquakes, for example, give rise to similar feelings of existential angst as people affected by them may for considerable periods of time struggle to find peace, fearing new aftershocks or similar disruptions of experienced safety, the totality of their impacts are delivered in a far shorter time span. The main difference in their phenomenology is thus the difference in the way their differing onsets affect lived experience.

5.2.2 *Fear of community disintegration*

One of the major recurring themes both during the expedition and in subsequent interviews and community-produced audiovisual content is leaders' concern regarding how the community might survive this hardship. With many having long since been displaced or moved elsewhere, for example to larger cities, it is unclear how many of these would return as part of the resettlement process. These worries centre on the questions of how many key personalities will have to remain for the community to 'survive' as well as how the recovery effort can ensure that the 'essence' of the community persists, at least in part. As such, one of the main anxieties connected to the disaster and the resettlement process appears to be the degree to which the community would persist throughout the process. This suggests that while being uprooted from parts of their ancestral land represents a major trauma in the collective memory of the Inga, the prospect of not persisting as a people is dreaded to an even greater extent.

Community representatives have stressed the potential of the displacement and uprooting to potentially adversely affect their cultural persistence. Having built their lives around strong communal values, the Inga thus fear the potential disruptive effect of displacement, losing their attachment to the land as well as potential conflicts that may arise as a result of the recovery process and potentially inadequate compensations. The risk of being scattered or losing certain community members who play an important role in maintaining social cohesion is a recurring source of worry. Losing neighbourly relations is another central concern. People rely on favours and reciprocal exchanges with neighbours, who, more often than not, represent a source of security and predictability for many, in combination with extended family. As one woman reports:

> The situation is that we do not have where to go but at least here with the neighbors I have lived well. I have lived comfortably for thirty years in this house but now I don't know what will happen. They say we may be sent away or something and this is what worries us all. We also have children. I have grandchildren who live here.

Others report similar concerns, underlining, in particular, the emotional toll separation from family and neighbours would have on people. It is clear that the community would experience less anxiety about the resettlement process ahead if they had faith that it would be people centred and focused not only on recovering the built environment but also on recovering the social aspects of the community, as much as possible. The worry is that if the town is reconstructed with a radically different plan and location of households, the social dynamic would not resemble the old town. This potential outcome seemed to be feared even more so than the disaster itself and could perhaps be seen as a potentially major secondary disaster to be mitigated.

5.2.3 Hope and distrust in authorities – centralisation dilemmas

The resettlement process and its ensuing intense feelings of uncertainty appears to be traced back mainly to a lack of trust in municipal and departmental authorities. While funding is a clear limitation to ideal planning outcomes, the Inga also express concern that public experts lack understanding of their cultural and social needs, critiquing the process for not having been sufficiently participatory. This is partially also shaped by prevailing views at the time of the study, where interim shelters that had been provided were of inadequate quality and were not regarded as suitable for a period of extended living while waiting for progress on the resettlement process. Anxieties connected to the potential for community disintegration thus revolved around the fear that the recovery would be only partial due to the risk of an improperly managed resettlement programme that would not be sensitive to social and cultural dimensions of recovery.

Several sources of distrust seemed to prevail, the first being connected to the quality of information. As mentioned previously, the experience of living in anticipation of an emerging disaster as well as an impending resettlement process generates intense feelings of uncertainty. Information on plans and next steps would then ease some of this uncertainty, or at least generate some predictability of what to expect in the coming months and years. For many of the Inga it was clear that anxieties also centred on whether their quality of life would be greatly reduced as part of the resettlement process. Not being able to envision the next steps thus became an extra burden.

A second source of distrust was connected to previous perceptions of neglect from departmental authorities. On several occasions, Inga representatives have expressed concern that public agencies have not had their interest in mind in the past, questioning the degree to which certain regional decision-makers have enough of an understanding of the needs of indigenous peoples and whether they respect their rights. On various occasions, it was suggested that involvement of the national authorities or international organisations would be preferred, perhaps as a way of minimising tensions that prevail among municipalities and local and subnational authorities. A sense of duty also prevailed as the efforts of the Inga in transitioning out of illicit crops was brought up. Since the people of Aponte had been on the side of the law, they expect in return that authorities come to their aid when they themselves face disaster. However, a sense of not being a high priority for the *Gobernación* in Pasto seemed to prevail among the local population, furthering the community's feelings of marginalisation.

A third source of distrust is thus connected to the lingering sense of not being a high priority for the relevant authorities. This is directly connected to previous concerns discussed, where the population is worried that the resettlement will be underfinanced and lagging, thus exacerbating the risk of a failed recovery and the potential deterioration of the community. Many community leaders, as well as members of affected households, reported feeling doubtful that the relevant authorities would invest sufficient resources and attention to their plight. Large parts of community-produced social media content is directed towards bringing attention to the situation in order to generate urgency. With a lack of perceived urgency being a common feature of slow-onset processes, the task of generating a sense of urgency requires conscious and persistent effort. Social media has in this way been used by indigenous leaders as well as youth in Aponte to document the disaster throughout its onset, powerfully demonstrating the common observation that people are not merely 'victims' of disaster, but can also actively resist adverse impacts. The relative capacity of the Inga, not only due to their connections with major political players, such as representatives of UN agencies, but also due to the political skill of their central leadership, have certainly proved integral to bringing some attention to the situation. However, the fear of an improperly funded and managed resettlement process still lingers, which is partially seen as a result of the unwillingness of subnational authorities to fully invest in the process.

Another source of distrust is connected to expert communication. As experts through the use of technical vocabulary have a tendency to 'render' everyday experience technical through the use of a mechanical logic, such oftentimes deliberate attempts at expert legitimation generates distrust on the part of affected households. In writing so I do not wish to imply that 'indigenous knowledge' is, in the eye of the Inga, in opposition to western instrumentalist logics as is often claimed— indeed, I was left with an impression that technical surveys were welcomed as these were central to achieving an understanding of the hazard phenomenon—but a crash does occur when experts fail to communicate accessibly. This connects back, in part, to my first point on communication, as a jargonised communication strategy serves to generate a sense of superiority among external visiting experts. Contrary to what is often thought to be the case, therefore, it appears to me that clashes of epistemology have more to do with conscious or unconscious acts of alienation on the part of experts than with 'indigenous knowledge' being distinct from 'scientific knowledge'. A dilemma thus occurs in trying to establish trust wherein technical experts may, on the one hand, risk conveying inaccessible or overly technical messages, or, on the other hand, conveying overly simplified messages which, in turn, produce a sense of alienation in the audience (feeling one is assumed to be ignorant).

To conclude, the issue of trust is closely connected to the topic of centralisation of responsibility and authority. Distrust among the Inga prevails mainly as a result of lingering feelings of neglect on the part of regional and municipal authorities. This distrust arises partially from having been ignored and alienated due to their status as a reserve, but also due to poor communication by some of the technical experts who have been involved in the recovery process. Feelings of distrust are likely to intensify if the community perceives that the salience of their plight is further reduced over time as the situation becomes protracted. For the Inga the stakes are clearly high, as the prospect of a failed recovery at the new site might cause irreparable damage to their long-term cultural survival and prosperity. The most dreaded outcome is perhaps that too many people will ultimately be displaced or otherwise not want to settle at the new site. For many, this anxiety is expressed not only as a fear of losing dear neighbours or important social and cultural assets, but also fear that the very future of their community is jeopardised.

5.3 Differentiated vulnerabilities

Central to subjective experience of disaster processes are structural and personal forms of vulnerability as well as capacities. If we see disaster as the material and non-material consequences of major disruptions of ordinary life and community functioning—often caused by a hazardous phenomenon coinciding with conditions of vulnerability, exposure and capacity—it makes sense to argue that the subjective lived experience of disaster will be substantially shaped by all of these factors, including vulnerability, capacity and exposure.[3] In more concise terms, we can also see disasters as circumstances that call for outside assistance to cope, whether at the individual or the societal level.[4]

Vulnerability can be understood as shaped by the totality of social circumstances. Individual traits and experiences, such as previous traumatic experiences, previous exposure to disasters, personality, and personal worldview as well as biological factors all play a role in how equipped an individual may be in the face— and in the aftermath—of disaster. Coping capacity can scarcely be assessed prior to actual disaster experience. Structural factors also play a significant role in the sense that one's position in class and power hierarchies not only shape individual personality, but also constrain (or enable) perceived or actual opportunities. In very simple terms, therefore, individual factors, on the one hand, and structural factors, on the other hand, shape and impact one another in ways that render disaster experience unique for every person, not only depending on their (seemingly) objective degrees of loss. This is not to say that people or communities are inherently vulnerable or not; outcomes are also shaped by prevailing capacities and their efficacy, as well as the intensity and type of the hazard. In the case of Aponte, for example, it is difficult to imagine how the community could possibly remain invulnerable to such a hazard, as its onset is perpetual and the disaster thus continually worsening over time. Still, individual and social factors produce different kinds of individual experience and loss, as will be the topic of the following.

5.3.1 Social ties and networks

The vulnerability of a household to disaster is partially shaped by its network and relationship to the community at large. Social capital, or the relative resourcefulness of members of one's close network and family, often serves as a lifeline in times of crisis. When losing one's home in a disaster, for example, having strong connections to others may be the difference between finding refuge with loved ones or having to rely on publicly provided shelter.

In Aponte, it is clear that social ties play a major role in augmenting disaster experience. Well-connected households were able to stay with family members or others in the cities, thereby contributing to rural–urban migration, as some are likely to remain even after ending of the resettlement process. Some drew on neighbours as a temporary solution while others had been forced to move to shelters. Many reported dread at the prospect of being moved to shelters, as the conditions were said to be inadequate. It became clear, therefore, that people valued their network strongly, and that seeing their social ties weakened as a result of the process was a recurring cause for anxiety. While this was not stated explicitly, one can also assume that leaders fear losing their relatively higher status in the community as a result of social changes resulting from the resettlement process and the perceived quality of its management. Whilst inherently dependent on social class, one can generally say that less well-off individuals are more reliant on networks to cope, but also have less resourceful networks to draw on. More influential individuals have larger networks, but are also less likely to be dependent on them for coping with disaster. This dilemma is a major part of how disaster experience connects to status and connections.

5.3.2 The campesinos

One large group of people living in and near Aponte that have not yet been mentioned in this book are the farmers, or *campesinos*, who are not considered part of the Inga. Many non-Inga have lived in Aponte for some time and moved there more recently due to the relative prosperity of the town following its many successes in generating social betterment for its people.

The disaster has left no one untouched. There is now a strong community sense of all being 'in it together'. Inga representatives, as well as municipal leaders, frequently reference the needs and condition of local farmers both in and around the town. While some of the farmers living nearby but not in the area are unaffected in the material sense, they still consider their stake in the process as the town represents their nearest urban settlement. Having maintained good relations for many years, many smaller farming communities are following the situation closely.

While the slow disaster affects everyone to a lesser or greater extent, not everyone enjoys the same legal status in the face of the adversity. Having a special legal status as a result of being a reserve, the relocation of Aponte involves special considerations relating to their cultural rights. As such, the resettlement discussions have split the community somewhat by making a distinction between Inga and non-Inga, at least at its initial stages. It appears as though this split is not attributable to a desire to stir conflict from either side, but rather that the official response had struggled to reach an agreement that secured both cultural rights and accounted for the needs of farmers. Either way, the extra uncertainty faced by the *campesinos* in the process was a source of considerable distress for those involved, even though the anxiety connected to cultural survival was not brought forward in the same way.

Towards the end of the data collection for this book, a new plan that better incorporated the needs of farmers was underway. While it is outside the scope of this book to speculate on how the resettlement process will ultimately turn out, it seems clear that the growing sentiment among *campesinos* that they have been overlooked in the process may linger for some time. This aspect of the resettlement process also forces us to take a critical look at the concept of community, which has also been argued for elsewhere[5] (although I will stick to using the concept in this book for lack of a better alternative). Guaranteeing the cultural rights of the Inga while also meeting the expectations of other affected populations has thus proved difficult for authorities involved in the resettlement process.

The farmers have considered themselves a part of Aponte and its socio-economic life. While perhaps participating to a lesser degree in cultural festivities and rituals, the *campesinos*, many of whom have migrated from other urban or rural areas, seem to find life in Aponte more peaceful and prosperous than where they came from originally. Some inhabitants had even moved in from Pasto, the departmental capital, such as one mechanic who had opened a garage in Aponte. Reflecting on his pleasure at living in a place where there was, for a long time, a strong a sense of progress and optimism towards the future, he is saddened to now see this future-oriented outlook being increasingly supplanted by pessimism. Because he has no strong cultural roots connecting him to Aponte, he stresses that he is not bound to the place in the same way as others, and that he would thus consider moving if his initial reasons for liking life in Aponte did not endure throughout the resettlement process.

As we have seen, the ways in which people are rendered susceptible to disaster and the way in which their experience and situation is shaped by it is largely dependent on their pre-existing circumstances. These include access to networks, legal privileges, income, healthcare, education and other factors which determine social standing and the ability to shape one's life circumstances. While vulnerability does indeed play a major role in shaping disaster experience and ability to cope, it is not as though people facing disaster are helpless in the face of adversities. People draw actively on their strengths, their networks and their skill sets to offset impacts to the best of their abilities. Many succeed in minimising disruption whereas others find their ways of coping inadequate to offset the consequences of disaster. In the next section, we will take a look at sources of resilience, where I will particularly focus on strategies drawn on to lessen the chance of a negative outcome from the disaster.

5.4 Capacities

Capacity is often taken to mean the sum of attributes, sources of strength and ways of coping that communities rely on in the face of adversities, such as disasters.[6] Towards the beginning of this book, I have strived to highlight how the Inga by no means appear as a 'vulnerable people'. On the contrary, the community is resourceful and maintains strong political and cultural influence across various levels, even to some extent at the global level, as represented by the Equatorial Prize award in 2015. As discussed in this chapter, we have also seen how Aponte has attracted people from all over the department of Nariño as a result of the quality of their schools and other social achievements. A strong sense of community has thus sparked sentiments of refusing to allow the contingency they now face to disperse and break their hard-won achievements. In some ways, the disaster is unifying, while, of course, also being experienced as a great tragedy.

Place-boundedness and a determination to care for the territory at any cost is often brought up as a motivation for facing the disaster head-on. In combination with a refusal to allow the community to disperse and the cultural heritage to fade, place-boundedness stands out as another source of determination. A common narrative is that worse challenges have been overcome in the past, including the occupation of guerrilla soldiers and paramilitary groups. As Wuasikamas, it is said, there exists a responsibility to care for the territory and the Earth as their forefathers have. Displacement thus constitutes a loss of meaning and the prospect of dispersion becomes a sort of ontological crisis where existential meaning would be challenged if both territory and community cease to be. Representatives of affected households and community leaders alike bring up the resilient nature of their people and a commitment to face the present challenge with the same stoicism that got them through past adversities.

I would also like to argue here that the tendency for the Inga to conceptualise the disaster as a trying period in their more extended history is a significant source of hope. Adopting a worldview in which current events are normally interpreted according to a long-time narrative perspective not only helps to distance oneself from present hardships, but also adds a collective sense of determination to overcome trying moments. Such a temporal perspective, in part based on a sense of living history, can thus be seen as a way to render present crises as parts of a historical legacy. This sense of witnessing collective history in the making arguably adds meaning to the lived experience of disaster, but, of course, does not offset the feeling of loss in the present. After all, life is mainly 'lived' in the present and immediate past.[7]

5.5 Living through and in the face of slow calamity

As I argue throughout this book, the defining feature of slowly manifesting disaster impacts from the perspective of lived experience is the way in which they trigger a sense of dread and anticipation connected to how they will continue to advance in the future. Will the disaster continue to be suffered in a more-or-less linear way, advancing with each passing day about as much as the last? Or will circumstances deteriorate exponentially and unpredictability? Or will this happen only after reaching some unknown future tipping point? Such future uncertainty-driven anxieties, I argue, are unique to the phenomenology of slow calamities. In saying so I do not intend to claim that disaster impacts suffered more instantaneously do not have a temporal dimension—as stated in the introductory chapter, all disasters have centuries of history serving as their precursors and their aftermath may forever shape the course of history. Yet the lived experience of the onset is clearly shaped by the speed of onset, not only of the natural hazard phenomenon itself but also the speed and duration at which its disastrous societal impacts manifest.

   The first sign of this is evident in the slow manifestation of the material damage to the built environment, in the form of gradually expanding and worsening 'cracks'. Affected households clearly expressed the anxiety connected to living with this phenomenon over an extended period of time, seeing a gradual deterioration week by week. This is not to say that the experience is more or less trying than the lived experience of going through a disaster that is experienced in a more immediately threatening way. Rather, the experiential aspect of slow calamities are distinct, and are, as I argue in this book, characterised by a sense of 'living in anticipation'.

   Upon noticing cracks in streets, sidewalks and the foundations of homes it seems clear that people initially did not foresee the extent of the problem. Only after the cracks were understood to be worsening and multiplying did it become apparent that something was terribly amiss. The onset thus increasingly dominated life and lived experience by becoming the main topic of conversation and the main issue of concern. Only after its potential consequences were felt and observed in the form of collapsing homes and displacement, did the experience of the phenomenon transition towards what we may label a disaster. In many ways living with the certainty that the calamity would continue to worsen and eventually consume the community was experienced as a secondary traumatic effect.

   The sense of living in anticipation, or seeing daily life in the present dominated by future dread due to current 'signals of worsening conditions', assumed many forms and was connected to different kinds of potential outcomes. Witnessing the slow disintegration of one's own home was certainly a major source of stress, but, as mentioned, the expected cultural, social and economic consequences were also feared.

   The prospect of being forced to resettle was perhaps the most immediate target of dread, even more than the material destruction suffered personally and collectively per se. Poor information concerning the resettlement plan caused people to feel that the intense uncertainty concerning to their individual and collective futures were all-encompassing. Most likely due to the fact that shelters were reportedly of low standard and without adequate access to hygienic facilities at the temporary site, people did not demonstrate faith in a successful relocation outcome. Rumours surrounding the inadequate shelters in this way increased the emotional hardship connected to the impending displacement of many households, especially those with children.

   Households with children and those taking care of the sick or family members with disabilities were particularly anxious about the situation. This can also be witnessed in a number of TV interviews from 2016, as this group of households have been particularly outspoken about the situation facing them. Such households worry both about how they will be able to continue providing care for their loved ones, and also about their extra need for clean water and other hygienic facilities at the temporary shelter. Of course, that the new homes constructed as part of the recovery process will be accessible for those in need of care is also a central concern. Families are also concerned about not allowing their children to fall into deeper poverty. Prolonged disruption of education is also a major concern, with the school building being one of the first buildings to suffer structural damage to an extent where it could not be responsibly used for teaching. Tents were therefore put up, but the quality of education and its frequency has varied. During fieldwork, it was also frequently brought up that people hoped the implementation of the resettlement process would not disrupt education excessively. Minimising impacts on children in particular was thus brought up as a priority for most families. Living with the certainty that one would be forced to resettle, but not knowing when and not knowing what kind of life one would settle into was a prevalent source of stress.

Life in anticipation, whether it be of the continued disaster onset or of the resettlement which came in its wake, is thus characterised by dread arising from a mix of certainty and uncertainty. On the one hand, the Inga knew with certainty that Aponte as they knew it would become completely destroyed and uninhabitable by most standards. On the other hand, they lacked certainty as to how this would play out and how the situation would be managed. We can say that in a sense the present becomes an irrelevant time from the perspective of lived experience, as the present is simply dominated by an expected future outcome. Anticipation of this outcome is therefore what ultimately ends up shaping everyday existence.

5.6 Problems with the concepts of resilience and recovery

The Inga of Aponte display many characteristics commonly associated with resilience or robust communities. Not only do they have an unusually inclusive governance structure and a tradition for participatory decision-making, but they have also, as we have seen, successfully resisted and stood up against guerrilla groups and paramilitary actors throughout the conflict in Colombia. The Inga have also been successful in eradicating illicit crops from their territory, opting instead for the cultivation of coffee. Many external and internal obstacles had to be overcome to see these changes through. These are a testament to the strong leadership on the part of Inga leaders.

The success story would later become central reasons for the Inga later being awarded the Equator Prize in 2015. The ability to frame their community in this way further demonstrates their political capacity to engage actively in national and global arenas for indigenous politics and activism. Differently put, although the Inga are a minority in the department of Nariño they do indeed defy what we may call conceptions of marginalised or vulnerable people based on the agency and persistence they have demonstrated in forming their own destiny and exercising their autonomy over a period of centuries. Therefore, from an internal or community perspective, it makes sense to conceptualise the Inga as a relatively resilient community. However, it remains unclear whether the concept of resilience can be usefully applied at all in such contexts where the community experiences pressing external challenges that are not easily overcome despite inward robustness. One such external challenge is, of course, the natural hazard phenomenon and the impending resettlement process, over which the community has little control. The phenomenon forces us to question whether the idea of being resilient in the face of a perpetually worsening hazard phenomenon is even worse, given that its intensity will not dissipate but simply keep worsening over time. Another question that arises is the implications of the local distrust towards wider society, particularly departmental authorities. In other words, can a community be considered resilient in isolation from its external societal context? Of course, the concept of resilience has been much discussed and critiqued elsewhere,[8] but these stand out as some additional inconsistencies within the concept.

The scholarship on disasters and resilience suggests that communities facing hazards will tend to transform, adapt or eventually collapse.[9] Often based on case studies of more rapidly manifesting disasters, prevailing models of resilience and vulnerability suggest that disasters tend to cascade and reoccur, exposing people to a potential negative vulnerability spiral which eventually exhausts local strategies for coping. How this dynamic functions in the case of more gradually manifesting disasters has not been sufficiently explored, with a few exceptions.[10] Reflections and insights based on the Aponte calamity give rise to the question of how we may understand the dynamics of resilience in the context of escalating and constantly worsening disaster impacts—impacts that eventually force communities to resettle elsewhere—which, in turn, raises the question of to what degree relocation can be conceptualised as a resilient move and, if so, under what conditions?

Research on post-disaster resettlement almost universally highlights that relocation should be seen as a strategy of last resort because of its high-stakes, low success rate and generally devastating socio-cultural consequences.[11] In discussions of the community resilience concept an underexplored, yet central issue is whether resettlement as a last resort should be seen as the community having collapsed or whether it can be understood as a form of recovery. When faced with the prospect of temporary shelter and impending relocation, many Inga chose to settle elsewhere, such as in larger urban areas. This dispersion of community members, in turn, potentially jeopardises a successful recovery, especially if central personalities are displaced and choose not to return to partake in the recovery effort, in the socio-cultural sense of the word, rather than merely in terms of reconstructing the built environment. In Aponte, key community figures have stated that they refuse to resettle and will remain in old Aponte even if it puts them at risk. This is grounded in a desire to protect the territory and live up to *Wuasikamas* ideals. As explained earlier in this chapter, many Inga have already been dispersed across the country and now reside in cities like Bogotá, Pasto, Popayán or Cali.

So does resettlement constitute a form of collapse (failure to adapt or reduce disaster risk), or can it in some instances be thought of as a form of recovery (if other options are unavailable)? This question is not easy to answer, but it has important implications for how we think about resettlement, recovery and the much-debated concept of resilience. Although a growing body of scholarship on post-disaster relocation teaches us that in some cases resettlement is a necessary last resort response, the literature is also very clear on its potentially devastating consequences, especially when it is poorly planned. Projected climatic changes are expected to further increase the relevance of resettlement as a strategy in some cases, mainly due to the fact that human settlements have historically been built in locations overly exposed to (increasing) disaster risk. One thing that seems clear is that resilience building and disaster risk reduction are not always available options; some areas are simply exposed to too great a disaster risk. Aponte, for instance, as we have seen, was destroyed by a slow-onset natural phenomenon that perpetually worsened over time, indicating that adaptation or 'resilience building' in situ was not a viable strategy in this context. It seemed clear already from an early stage that the perpetually worsening disaster would not dissipate——the destruction would continue unchecked until both the whole terrain and everything built atop it was completely destroyed. Community resilience in the way we normally understand it thus seems like an awkward concept for describing these kinds of scenarios and phenomena. Differently put, the literature on resilience and adaptation often makes the critical error of assuming, either implicitly or explicitly, that withstanding or being able to bounce back following a disaster is always possible. However, in the case of perpetually worsening hazards and their slowly accumulating disastrous consequences, notions of resilience fall particularly short, at least if it is understood in terms of enduring in a specific place.[12] The same points may be raised for the concept of recovery.

Depending on how the resettlement process evolves it seems possible that the resettlement process could give rise to a split, with a new town consisting of relocated households as well as an old town amidst ruins where some people keep on living and others return. Such a tendency for old and new towns resulting from resettlement processes is not uncommon, as witnessed from numerous case studies from other contexts.[13] Such an outcome is not necessarily negative as it preserves a connection between the new and the old, although it does potentially expose the remaining population to continued disaster risk. Additionally, if Aponte is eventually fully abandoned, with many of the displaced people returning for the community recovery effort at the new site, it may be possible to recreate a similar community structure. Ultimately, it will be the community itself that judges whether they experience having survived the move or not, as in whether they have been able to re-establish a similar socio-cultural dynamic and a sense of continuity despite a geographical relocation. It remains unclear how changed or altered the new town can be before it will be experienced as an altogether different community by its inhabitants. This has implications for whether the resettlement process is accepted by its stakeholders as a failure or a successful case of 'recovering elsewhere'.

Notes

1 Interview coverage with Pasto Noticias May 5, 2017.
2 Neal (1997).
3 Inspired by UNDRR's definition of disaster, see: https://www.undrr.org/terminology/disaster
4 Kelman (2020).
5 Titz, Cannon and Krüger (2018).
6 Inspired by UNDRR's definition of capacity, see: https://www.undrr.org/terminology/capacity
7 Adams, Murphy and Clarke (2009).
8 I am not citing any particular article here, but a large number of publications have criticised the concept, although not necessarily rejected it. See, for example: Weichselgartner and Kelman (2015); Manyena (2006); Klein, Nicholls and Thomalla (2003).
9 See, for example: Pelling, O'Brien and Matyas (2015).
10 For an overview, see Staupe-Delgado (2019b).
11 See, for example: Oliver-Smith (1991).
12 It must be noted that it might have been possible to find engineered solutions, but there are likely to have been too costly and also it is unclear whether they could permanently solve the problem.
13 See for example: Oliver-Smith (1979).

6 The ancestral land

Territory, community and resettlement

Much has been written on the importance of territory and the strong place-boundedness of many indigenous communities. Indigenous communities are often described as 'rich and diverse cultures based on a profound spiritual relationship with their land and natural resources'.[1] This, in turn, may be traced back to arguments stressing that distinctions 'such as nature vs. culture do not exist in indigenous societies'[2] because they 'do not see themselves as outside the realm of nature, but as part of nature'.[3] They are thus said to 'have their own specific attachment to their land and territory and their own specific modes of production based on a unique knowledge of their environment'.[4] Further, indigenous communities are often described as grounded in collectivist cultural traits, as having distinct worldviews, ways of knowing as well as practices. However, portraying indigenous peoples in this way risks generalising and simplifying a more complex picture on the grounds of problematic stereotypes. In reality, indigenous communities and persons are diverse and hold varied ambitions and views of past, present and future. Yet there are also aspects to indigeneity that are distinct, although we ought to also recognise that individuals vary in the extent to which they emphasise their indigeneity and how they orient politically to assert their rights.

A common theme is that indigenous peoples have strived to preserve their cultural heritage, or parts of it, through the formation of strong traditions aimed at keeping alive their expressions and beliefs. These include art forms, musical styles, dances and other forms of cultural beliefs, spiritual practices as well as traditional rituals and forms of symbolism. The means and ends for the preservation of heritage vary from a desire to protect some essential cultural expressions over time, to other communities who emphasise a traditionalist way of life. Contrary to common myths, however, the majority of indigenous communities and individuals do desire to adopt technologies that not only make life more convenient, but also facilitate cultural expression, such as social media and digital technologies. In other words, only a minority of indigenous communities actively strive to isolate from global affairs, meaning that many communities constantly renegotiate how to balance their aim of preserving cultural expressions with their desire to partake and contribute to global technological developments.

Ensuring their cultural prosperity is perhaps most essentially seen as an emphasis on the right to speak and pass on their own dialects and languages. The development of a written language has been a central part of this process in the case for

DOI: 10.4324/9780429288135-8

languages that lacked it. Yet the role of territory cannot be downplayed. Ancestral lands and the connection between people and place remain central to the way in which indigenous communities have traditionally exercised their rights, in addition to the use of one's native language.

Territory and land plays a central role to the cultural identity and political organisation of most indigenous communities. Therefore, in cases of displaced indigenous populations the connection to some now lost but former ancestral land also tends to be strong. In places where legal rights to territory are not officially recognised, indigenous communities often centre their political agency on securing formal connections to previously owned lands. Territory often forms the basis for traditional economic practices, such as animal herding, hunting or traditional handicrafts—and often function as political centres even for members who are not living in the territory per se. However, the centrality of territory generally extends beyond the political and economic sphere. It is not uncommon for indigenous political leaders to emphasise a strong place-boundedness rooted also in spiritual practices and a sense of belonging. Identity is often tied up to local flora and fauna, meaning that some practices cannot necessarily be practiced elsewhere. Local place names will often have names in the indigenous language, further contributing to a sense of belonging and recognition. Families also often take the name of local areas where they and their ancestors lived, such as the name of a farm, a hill, a valley or a river, sometimes in ancient versions of languages now modernised. Moreover, the territory is often where ancestors have been buried in centuries past and thus remains important for the lived experience of historical continuity or being in the world. The centrality of territory is also explicitly emphasised in a United Nations forum for indigenous issues:

> Land is the foundation of the lives and cultures of indigenous peoples all over the world. This is why the protection of their right to lands, territories and natural resources is a key demand of the international indigenous people's movement and of indigenous peoples and organizations everywhere. It is also clear that most local and national indigenous people's movements have emerged from the struggles against policies and actions that have undermined and discriminated against their customary land tenure and resource management systems, expropriated their lands, extracted their resources without their consent and led to their displacement and dispossession from their territories. Without access to and respect for their rights over their lands, territories and natural resources, the survival of indigenous peoples' particular distinct cultures is threatened. [...] Land rights, access to land and control over it and its resources are central to indigenous peoples throughout the world, and they depend on such rights and access for their material and cultural survival. In order to survive as distinct peoples, indigenous peoples and their communities need to be able to own, conserve and manage their territories, lands and resources.[5]

While many peoples have strongly constructed their culture and identity around territory and thus achieved a strong sense of place-boundedness, other groups have assumed more nomadic lifestyles. Indigeneity is often associated with the idea of original occupation, sometimes since 'time immemorial'—or otherwise understood to involve peoples displaced from their ancestral lands, or under constant threat from what may be labelled colonial processes or other kinds of existential or cultural threats posed by settler populations or political projects aimed at assimilation. However, such place-based forms of recognition would not allow us to consider nomadic peoples. Thus, indigeneity also centres strongly on ancestry and cultural practices. Nevertheless, it should be noted that as each legal system employs different ways of qualifying indigeneity and special rights from a legal point of view, formal recognition of indigeneity is also in part political. This also means that some individuals choose to pursue recognition, while others may prefer to avoid emphasising their indigeneity for a number of reasons. Such reasons may range from stigma, for example as a result of negative attitudes towards indigenous peoples in the general population, to a desire to disassociate, such as a young person experiencing discrimination within a community and wishing to disassociate from that community.

Land tenure and formal rights over territories as well as natural riches within these territories constitute some of the most central and most contested foci for advocacy groups and indigenous political leaders. Although progress has been made in this regard, many communities still fear further dispossession and perhaps in particular slower manifestations of empire in the form of cultural imperialism and an abandonment of traditional values and practices in favour of increasingly Western and consumerist lifestyles. Indeed, the commodification of indigenous culture and products has been raised as an issue in many fora of indigenous politics. However, younger voices in particular also express the need for cultural dynamism and new forms of cultural expression, involving also social media and a stronger global presence. Some communities have in this way succeeded in building a strong community identity in digital spaces on the internet, which some (controversially) argue may represent an important strategy for cultural survival in a hyper-connected age. In Aponte, we also see this strategy being used as community leaders move their activism online to secure greater support not only nationally, but also internationally.

A central legal instrument for indigenous politics is the United Nations Declaration on the Rights of Indigenous Peoples, which also explicitly notes the importance of territory for the cultural identity of most indigenous communities. Its Article 26, as seen in the box below, emphasises the importance of formalising traditional forms of tenure to secure ownership, as well as the right to autonomy and protection of customs. Also of relevance to this book is its Article 10, emphasising the importance for resettlement to be inclusive and voluntary, in those circumstances where no other options are available. Forcibly removing a population from their ancestral lands, therefore, means breaking these normative principles, even if it is carried out with the intention of reducing the population's disaster risk.

Article 26

1. Indigenous peoples have the right to the lands, territories and resources which they have traditionally owned, occupied or otherwise used or acquired.
2. Indigenous peoples have the right to own, use, develop and control the lands, territories and resources that they possess by reason of traditional ownership or other traditional occupation or use, as well as those which they have otherwise acquired.
3. States shall give legal recognition and protection to these lands, territories and resources. Such recognition shall be conducted with due respect to the customs, traditions and land tenure systems of the indigenous peoples concerned.

Article 10

Indigenous peoples shall not be forcibly removed from their lands or territories. No relocation shall take place without the free, prior and informed consent of the indigenous peoples concerned and after agreement on just and fair compensation and, where possible, with the option of return.

Source: United Nations Declaration on the Rights of Indigenous Peoples.

As described in the examples above, the connection between peoples and territory are also best understood as nuanced. Individuals vary in terms of how strongly they identify with their broader community, and some members of a community may end up wishing to disassociate for different reasons. The contested nature of the community concept thus suggests that community is a great resource for many, but a sort of tyranny for some. Communities also vary in inclusiveness or toleration for difference; some are more conformist than others. Therefore, people who find themselves unable to conform to community expectations may find little personal security derived from community and instead wish for protection from the community. For marginalised minorities, for example, the broader community may be their greatest source of fear and risk. LGBTQ people are but one example of a group that may be at risk of community exclusion depending on the values and structure of the community in question.

Indeed, typical accounts of the connection between place (or nature) and indigenous peoples as 'more-than-human'[6] end up leaving more nuanced discussions out of the picture. Such accounts also overlooks important historical and political struggles of recognition and exclusion. To this day, who may or may not be legally recognised as indigenous or not within countries remains contested. Furthermore, some communities may be more exclusive than others, preferring to not take in members who (have) lead non-traditional lives for some time. The role of the mother tongue is also a matter of debate and has been used to exclude people who want to pursue recognition. Conceptually, therefore, 'indigeneity attends to the social, cultural, economic, political, intuitional, and epistemic processes through which the meaning of being Indigenous in a particular time and place is constructed'.[7] To further problematise the concept a parallel may be drawn to cartography:

> Like maps, indigeneity also functions as a style and manner of representing the outcomes of specific historical and geographical processes as facts, naturalizing the asymmetries of power characteristic of colonialism through the assertion of an essential connection between place and identity ... Indigeneity works as a residual category, referring to everything that existed prior to all that is Western or modern[8]

The contestable nature of the question of who may assume indigenous status and who may not thus gives rise to a relatively flexible category of identity. Resourceful actors may have interest 'and means of fashioning indignity in their mould', but so too will individuals who disagree with dominant views within a community and find the concept necessarily flexible to open up for other forms of representation and activism.

In sum, the connection between peoples and territory is not straightforward. Territory surely plays an important role for the livelihoods and cultural identity of many indigenous communities. It is nevertheless important to recognise that indigenous communities, as well as their individual members, vary as much as people in general, both in their views and forms of representation. Equally importantly, as has been mentioned previously in this chapter, some community members may also find themselves subject to discrimination within their group, wishing instead to find protection from their community, as opposed to seeking protection within. In discussions of indigeneity, therefore, I believe it is important to avoid typical policy-level narratives treating indigeneity as a distinct category or something 'other'. While the role of territory and certain cultural practices may play a central role for many indigenous communities, including the Inga, the importance of both in everyday life varies from person to person.

6.1 Narratives on the importance of territory for the Inga

After arriving in the territory over 300 years ago, the Inga have remained attached to the land left behind by their founder, Chieftain Tamayo, who is said to be watching over the territory to this day. Having strived for many years to formalise tenure for their land and obtain the status of *resguardo*, based on the original lands declared as belonging to the people of Tamayo, the Inga have since adopted an identity of being carers for the land, or Wuasikamas.

As we have seen, the connection between people and land is multifaceted and involves a number of aspects, including emotional and instrumental ones. This has important implications for how we think about displacement and resettlement, as place plays an important role in the experienced sense of continuity as community and people. Among the Inga there appears to be a number of themes that illustrate the lived experience of place-boundedness and fears arising in the face of impending resettlement.

6.1.1 Ancestral land and spirituality

The ancestral land plays a major role in the *cosmovición* Inga. Both their history as a people as well as recent developments have shaped this special bond. This is perhaps owing to their recent history, culminating in the oath of Wuasikamas and the abandonment of poppy cultivation which strengthened this feeling of determination and belonging. The lived experience of relative timelessness is also frequently highlighted as meaningful, with many informants stressing how they personally find it spiritually significant to know that they live on the land of their forbearers. Two statements serve to illustrate this point:

> Imagine that the founders lived here centuries ago. This is why these lands are so important to us as a people.
> The presence of our forefathers in these lands as part of the *cosmovición* Inga also tie us to these lands.

While the territory itself is drawn on by the Inga for spiritual fulfilment, the hazard is also destroying much of the material aspects of their spiritual fundament. Shortly after my visit the local church in Aponte collapsed. Identifying as Catholic, but with a faith that also greatly draws on Andean traditions, worldviews and spiritual practices, the loss of their church represents a major loss for the community. Not only a place for worship, the church was also placed at the centre of the town square and was a place for the community to come together to connect spiritually, not only in the Catholic sense.

Mourning the loss of their place of worship, many people seemed to increasingly experience a sense of melancholic placelessness when looking at the ruins (or gradual ruination) that were once their home. Even if a potential relocation site would likely be available not far from old Aponte, a sense of disrupted living-in-place would be likely. The loss of particular sites, such as churches and community buildings, as well as the town square, are experienced as emotionally difficult and tragic, not only for the direct material loss but also because of how it affects the lived experience of historical coherence in a place.

6.1.2 Home, neighbours and family

People connect meaning to homes (the experience of feeling 'at home'), clusters of neighbours and friends in relation to one's own home, as well as the presence of family and relatives. The spatial layout of the built environment in this way also shapes the dynamics of social interactions and relationships. While neighbours may have their differences, having lived together with a certain proximity to one another shapes social life in significant ways, including the reciprocal exchange of favours and the security this offers. Securing the survival of certain social dynamics and locally valued aspects of community life will therefore be an essential part of the resettlement effort.

As previously mentioned, it may be useful to think about resettlement in terms of recovering elsewhere. The concept of recovery should in this context be understood broadly, also involving social, cultural and livelihood aspects of recovery. Reducing recovery to a focus on the built environment would in this sense be simplistic and potentially threaten community survival in the aftermath. In this way, neighbourhoods and the reconstruction of these are essentially about spatial aspects of social relationships, again affecting how life is lived in a place which, in turn, shapes culture. For example, if neighbour A frequently interacted with neighbours B and C (e.g. initiating the organising of community events) it seems likely to assume that the rate of community events organised would change if the personalities of person A's new neighbours were different (less willing to join in organising community events initiated by person A). Another example could be cases in which a person relies on neighbourly services, such as sharing caretaking work for children or the elderly. A third example may be if a person in need of care depends on neighbours to carry out basic tasks such as grocery shopping or even just social company (having caring neighbours who regularly comes over to 'check in' was cited as important for many residents). Ensuring that as many such relationships survive would thus be an essential component of the recovery effort if the goal is for the community dynamic to survive.

The nature of housing is also a frequent source of dissatisfaction following both reconstruction efforts and resettlement schemes. Companies or actors charged with building the new homes and buildings will often tend to do so according to a plan that lacks local ownership. Outside actors may base their architecture on non-traditional designs, leading to homes lacking certain aspects seen as essential, such as a patio big enough for washing and drying clothes. Standardisation of home layout and standard is another contested issue. Providing every family with the same type of home does provide a sense of fairness from a limited perspective of justice (e.g. a standard model home with two bedrooms, kitchen, living room and patio). However, a larger family which had four bedrooms would find their quality of life significantly reduced if they were given a two-bedroom house as a replacement. Moreover, because diversity in home layout arguably reflects diversity in needs, the needs of many families would not be met if the type of home they had originally is not taken into account.

6.2 Recovery as a slow healing process centred on community rehabilitation

Scholarship on post-disaster shelter and longer-term recovery has brought up the centrality of long-term perspectives in such processes. Arguing for the need to understand recovery in socio-cultural terms and as a form of slow healing, Kasia Mika and Ilan Kelman, among others, propose a processual view of disaster recovery.[9] A fundamental assumption underpinning popular notions of recovery is the need for returning to normal in a swift and effective manner. Indeed, most response agencies emphasise effectiveness as a yardstick when assessing their work. Recovery is in this way seen as being mainly about recovering lost buildings and infrastructure. However, and as emphasised in this work, rebuilding social aspects of community takes time and effort, just as overcoming traumas often represent a lifelong challenge.

An appropriate starting point is a recognition that it is unclear whether the loss, destruction and uprooting from disasters can ever be recovered from in the first place, or if normal conditions can realistically be completely restored.[10] It should also be added that it is not certain that pre-disaster conditions are not necessarily something one would want to reproduce. Ultimately, the existence of privileged and underprivileged social standings (which also usually shape the disaster outcomes of households and individuals) give rise to a skewed balance of impacts.[11] We would therefore expect that some would want to restore normalcy whereas others would like to see changes to the status quo as part of the recovery effort.

The emphasis on slower and more processual forms of recovery, as presented by Mika and Kelman, emphasises the emotional work and effort involved in survivorship and displacement and the steps involved in rebuilding communities and lives. Based on such a conceptualisation, then, recovery is also seen as having a strongly affective dimension which is focused not only on reconstruction but also on the grief, melancholy and sense of having to recover from loss that also characterises disaster aftermaths after immediate material needs are covered. Recovery thus involves coming to terms, but also requires action in the form of taking charge of one's future, and exercising one's agency to contribute to long-term holistic recovery. In this way, a conceptualisation of recovery emphasising its slow nature aims to:

> capture the slow, yet no less urgent, process of rebuilding, plodding towards, and climbing up to a shelter for oneself or a cluster within which a disaster survivor could dwell, such that s/he can, once again, imagine possible futures, from relationships of care, and forge one's own mode of being and belonging in the world.[12]

As we can read from the excerpt above, the psychosocial aspects of recovery involve in a sense moving on and reorienting one's place in the world. Nowhere is this more evidently a big part of the recovery process than in the context of displacement and resettlement. In recovery processes involving community relocation, the spatial reorientation that will be demanded of people will also involve rebuilding potentially severed social ties, reorienting oneself not only in a place but also in a neighbourhood and village context. Livelihoods may have to be reassessed and individuals, households and entire communities will necessarily also be challenged to reform their social dynamic. After all, nobody possesses the foresight that would be necessary to predict how social and cultural change would evolve over time as a result of the disaster, much less how the disaster will shape the history of a community or a people.

6.2.1 Territorial aspects of the resettlement process

The economic and socio-cultural uprooting associated with resettlement can be a source of collective trauma. At worst, resettlement can result in community collapse, witnessed as the dispersion of community members and abandoning of previous social ties. At best, resettlement can be understood as an emergency recovery measure of last resort that can, if done correctly, reduce disaster risk and provide a valuable opportunity to reshape collective identities. Highly successful examples are hard to find, however. One determinant appears to be how far the new settlement is from the old. Another is the degree of local involvement in the process. Experience tells us that geographically extensive relocation processes that disrupt livelihoods hold little potential for the betterment of affected peoples' lives. The same may be said for centralised resettlement initiatives that fail to include affected populations, leading to socially, culturally and economically nonviable conditions at the new location.

Since the beginning, an important goal in the Aponte resettlement process has been to either find land nearby or to find suitable land within the existing territory. Since most of the current settlement is also exposed to significant future disaster risk, most homes cannot be reconstructed in the same spot. Already, by the beginning of 2016 a geological survey was initiated which aimed to identify the extent of the disaster risk exposure in the area and to identify which areas of Aponte could remain and which would have to be resettled and demolished. Because of bureaucratic delays as well as a lengthy investigation into the matter, the process of mapping disaster risks and vulnerabilities dragged on, leaving displaced families in improvised shelters and poor housing conditions. For some families, this meant living in precarious conditions for years. Formally, the families had been promised and many had been given rental subsidies, but a lack of available housing for rent in the region made this an unsuitable solution for many. This was particularly unsuitable for larger families who would need to rent an entire house.

Once the survey was finished it was determined that only a small portion of the Aponte territory was suitable for the recovery effort. Having lost three-quarters of the built environment to the geological fault, thus having just a quarter of the town standing relatively unscratched, it was clear that a lot of land would be needed to reconstruct 75 per cent of the town at an alternative site. The survey eventually determined that four hectares, representing a sixfold reduction in the size of land, seemed suitable for permanent settlement. In other words, great uncertainty was connected to how the community could prosper within such confines. With over five hundred families waiting for urgent recovery of homes at the time, then living in shelters and temporary housing, solutions would have to come sooner rather than later.

One question that arose was how to balance the delicate dilemma of whether it is best to resettle within the boundaries of the existing territory, or perhaps identify alternative sites. Resettling within the territory on those four hectares, or alternatively somehow expanding these to a slightly bigger area by buying up some more of the surrounding land, is certainly the most attractive option and the one ideally preferred by the Inga and decision-makers alike. However, the crowding that would ensue is a problem, leaving Inga leaders with few options. On the one hand, settling within the territory would minimise the potential for socio-cultural disruption associated with resettlement. On the other hand, a territory larger than the four hectares deemed safe in current Aponte could potentially be found elsewhere. This dilemma had not been settled at the time of writing as the final decisions and initiation of the resettlement process became far more protracted than anyone could have anticipated.

6.2.2 A long process

Physically recovering the built environment from disaster often takes a very long time. The social aspects of recovery often take even longer, and oftentimes taking the time to recover durably may be better than swiftly recovering to an undesirable state. Once it became clear in early 2016 that relocation would be necessary for well over three-quarters of the population of Aponte, no one was in a position to foresee exactly how things would play out. Yet, having experienced marginalisation in the past, the Inga were clear on that unless a very sustained effort was maintained at getting the attention of decision-makers, including representatives of international organisations, their plight would most likely pass unnoticed. The Inga knew from an early stage that they would be forced to mobilise in every way for a very long time to keep their recovery on otherwise busy political agendas. Indeed, new challenges kept emerging, pushing the Aponte disaster further and further down on the political agenda. In the end, the inhabitants of Aponte are still awaiting concrete steps being taken towards resettlement and recovery at the time of writing this chapter. With the fear of a failed resettlement looming large, and exhausted from years of living in increasingly precarious temporary housing, many members of the Inga look towards the future with great anxiety.

The situation is not made better by the COVID-19 pandemic, which is ongoing as this manuscript is being submitted. For the first two months of the pandemic it appeared as if Colombia had managed to keep cases low, although testing was initially a problem, of course. As of the latest revision of this chapter, over 60,000 people have died from the disease in Colombia. With numerous and recurring lockdowns across the country's major cities, the political salience of existing plights quickly dropped. Surveying the news landscape, it is clear that survivors of recent disasters such as the Mocoa landslide and the Aponte disaster are not mentioned, although the recovery of both remains elusive. At this point in time it is impossible to know how the resettlement (and hopefully the recovery) of Aponte will play out in the future or if this will become a forgotten disaster.[13] Having promised to tell their story to the best of my abilities, I hope that this book delivers on this objective, while also having clear analytical implications for disasters in other contexts.

Notes

1 UN (2009: 52).
2 Ibid.
3 Ibid.

 4 Ibid.
 5 Permanent forum indigenous peoples statement, cited in UN (2009: 54).
 6 Radcliffe (2017: 220).
 7 Ibid., p. 221
 8 Bryan (2009: 25).
 9 Mika and Kelman (2020).
10 Mika and Kelman (2020); Hills (1998).
11 Wisner et al. (2004).
12 Mika and Kelman (2020: 2).
13 See for example: Wisner and Gaillard (2009).

Part III

Reflections

7 Living in anticipation of impending calamity

Towards an analytical notion

This chapter sets out to describe an analytical notion I refer to as 'living in anticipation'. It refers to the way in which expected calamities ('disasters waiting to happen') or disaster impacts slowly unfolding (slow calamities) end up shaping—and potentially overshadowing—life in the present. Once the existence of disaster risk or an actual disaster onset is identified, their hitherto destructive potential becomes a matter of fact with implications for how life is lived. In other words, they become real through their ability to shape the daily life of exposed populations, even before their full destructive potential is felt. This happens in the absence of their actual occurrence, as is the case for a discovered disaster risk, and in the hitherto nonarrival of the 'peak' in the case of slow calamities already in progression. The process itself depends on noticing this potential, often with the help of technology and scientific inquiries, seemingly bringing the anticipated future consequences into the present through various forms of forecasting and prediction efforts, which, in turn, exert influence on life in the present. After all, life in the present is to a great extent shaped by ideas about the future.

The analytical heuristic is of relevance for a number of contemporary conditions that most readers should be in a position to relate to. At the time of writing, the world is shaped by the global COVID-19 pandemic. I invite readers to think back to the early onset of the pandemic when for most countries outside of China life seemed normal and relatively unaffected by the rumours of a novel strain of coronavirus affecting Hubei. Shortly after, seemingly overnight, the global pandemic threat had been established through forecasts and epidemiological modelling, causing many countries to implement measures of varying degrees of strictness in an effort to 'flatten the curve'. Life had thus gone overnight from being relatively normal to being shaped by anticipation or dread of the expected peak, which was predicted for May, in most European countries. Lived experience had been nearly completely overshadowed by the expectation that the peak could become too steep, that we might be on the wrong track, or that measures may perhaps prove insufficient to cushion the worst consequences and keep mortality to a minimum. As is no surprise in hindsight, countries varied widely in terms of how they fared and how they approached the pandemic. This discussion beyond the scope of this book. The point remains, life was for some time dominated by perceived disaster risk, and thus our experience was one of living in anticipation. Secondary effects played a similar role, as worry and speculation over whether economic collapse,

DOI: 10.4324/9780429288135-10

cultural collapse or social collapse was on the horizon caused fear and anxiety in ways that shaped or perhaps, in some cases, even dominated lived experience in the present. Phenomena as diverse as climate change, antimicrobial resistance, El Niño-Southern Oscillation (ENSO), earthquakes or environmental change hold the potential to create similar affective states in the sense that their onset allows for a temporal experience shaped by dread and seeming powerlessness to stop them (at least the hazard process, albeit not necessarily their consequences) from progressing. Another aspect to this experience is the realization that many elusive phenomena that produce a sense of living in anticipation may also be collectively actionable, but go unaddressed due to political obstacles, effectively leaving adversely affected communities with few opportunities to address root causes. This further exacerbates the sense of lock-in produced by such phenomena and their resulting sentiments, giving rise to feelings of hopelessness.

In the case of Aponte, the geological hazard it faces also rendered life increasingly marked by anticipation and dread as the slowly manifesting calamity advanced, week by week and month by month. The disaster was experienced as a locked-in future event that could not be stopped once its onset had begun progressing. This is not to say the disaster is natural or that its deadly potential is unstoppable. Indeed, from a technical point of view it is easy to get people out of harm's way when the destruction of the built environment happens at a slow pace. Still, the impacts of the ruination of homes, the disruption of livelihoods and the negative repercussions they have on the cultural survival of an indigenous reserve can itself be considered a disaster, despite few to no direct casualties attributable to the natural hazard phenomenon. A peculiar aspect of calamities of this type is that their severity first becomes apparent once serious investigations lead to a realization that their onset is not temporary, but that the mild destruction witnessed initially is but the early onset of a very long-lasting and destructive process. As the news sinks in and the disaster looms larger each passing day, life is increasingly dominated by dread and worry about the inevitable hardships that await. This affective state is the essence of what I refer to as living in anticipation.

7.1 Conceptualizing the lived experience of anticipation in the context of looming and slowly manifesting disasters

Among the central characteristics of our present time is an increased orientation towards governing time and futures, or what we may refer to as a 'politics of temporality'.[1] Based on a seeming contradiction between enduring aspects of social experience and rapid shifts or accelerations, capitalism and its technocratic underpinnings at once imply better prediction as well as a greater appreciation for uncertainty (or the limits of knowledge, given that the state of knowledge is always changing, evolving and sometimes collapsing). Prediction and the prevention of societal challenges is in this way often thought to be on the horizon, located in a not-quite-yet time——a destination at which we never quite seem to arrive, owing to the frustrating realization that our knowledge of future contingencies is still not as complete as we might have hoped. Better knowledge about the future is thus postponed 'as more sophisticated ways of knowing it continually emerge'. As such, the role of the future determines much of what we can consider current forms of societal organization. Although its role can hardly be considered new (all permanent settlements, by definition, had to base their organization around anticipation of dangers, harvests, weather and demographics), technological development has enabled the unravelling of hitherto unknown disaster risks and contingencies, with some notable limitations. In many ways, the experience of living in anticipation may be better described as dominated by a sense of certain uncertainty, than of accurate prediction. Both the experience of living through slow calamities already underway, as well as scheduled disasters, such as life in places exposed to extreme geological hazard risk, are therefore shaped by certainty of the existence of danger, but uncertainty as to its eventual aftermath. Their disaster risk potential remains open from the perspective of current day-to-day experience of disaster risk.

The search for increased predictive power and forecasting methodologies is at the heart of applied science. In a much-cited 2009 paper, *Anticipation: Technoscience, life, affect, temporality*, Vincanne Adams, Michelle Murphy and Adele E. Clarke theorize practices of anticipation and argue that this search for better prediction renders the future (and ways of anticipating it) 'always knowable in new ways, even as the grasping for certainty about it remains persistent'.[2] Critical reflections of how projected or even speculative futures are engaged in the present are uncommon, even among those who act on it for a living. Despite anticipation (discussed by the authors as a 'lived affect-state of daily life') being a pervasive mode of experience, cutting across such mundane aspects of our existence as how we understand ourselves, make sense of our lives, monitor our health and orient ourselves towards the natural environment as well as spiritually, this lack of active cognition around its implications for ourselves and society often remains true. In seeing the lived experience of anticipation as an affective mode, life in anticipation (of disaster and calamity) can be understood as a mix of 'needing to know'—a search for more reliable knowledge about potentialities—and a feeling of dread for an expected or creeping calamity. In other words, anticipation can be understood as a form of 'palpable effect of the speculative future on the present'.[3] Not merely reactive, Vincanne Adams and colleagues see anticipation more in terms of 'a way of actively orienting oneself temporally'.[4] Seen as an effective state, the lived experience of anticipation encompasses both neutral forms of uncertain event horizons (a 'cliff-hanger'), where the feeling is not necessarily one of dread but of suspense, as well as feelings of intense anxiety emerging from a realization that the future will be, or is likely to be, disastrous. In most cases, such a sense of impending hardship can be met with preparations to mitigate the risk; in other cases, however, there may be too little time or too few factors that can be influenced, forcing people to hope for the best, but to anticipate the worst.

In the context of disaster, the future can be conceived of as inherently risky. Despite our best efforts to detect, assess and mitigate disaster risk, residual risks may remain either in the form of yet undiscovered hazardousness, or as disaster risk not appropriately scaled. Political factors may also obscure these processes, for example, by deeming it inconvenient to mitigate the risk of truly great, but unlikely calamities. We may thus understand the future as continuously unfolding before our very eyes, as an extension of the present of sorts projected into the future. Indeed, the direction of present trends are often extended into the future without concern for its contingent nature——a fallacy often made in economic and social predictions. Analogous to a vessel at sea, the coast is more often than not clear (ages come and go with no disaster occurring at a particular place), with some familiar reefs we know well to evade, and some whose existence we have detected and in relation to which we have taken appropriate measures. A few remaining ones are unknown to us, however; our chances of hitting them are low and continuing at full speed ahead is generally perceived to be safe (after all, experience tells us that the general risk is low), but eventually we might just hit one hard. When new signals about what lies ahead become available, it would be wise to alter one's course accordingly. Living in anticipation is thus an active orientation towards a 'more-or-less expected future in which our "presents" are necessarily understood as contingent upon' a continuously evolving and adaptive future——a 'future that may or may not be known for certain, but still must be acted upon nonetheless'.[5] The implications of learning that Aponte was being affected not only by a geological hazard, but one that would worsen relentlessly, is an example of such an 'existential slap'.[6]

Preoccupations with the future and speculations and predictions about what it harbours (crises, contingencies, near-misses or otherwise) is, in combination with narratives about the past, a defining feature of lived experience in the present. Daily life, consisting of a series of tasks, errands and ordinary moments is in many ways a projection of what lies ahead, informed by anticipations and expectations of and for the future. Saving for periods of ill health, buying insurance, diversifying crops, applying pesticides, sowing seeds, attending school or even worrying about potential risk to oneself and one's family are all examples of future-oriented practices. Paradoxically, it is often only when everyday life is disturbed by contingencies that we desire the routines of our previously uneventful lives and long for a time when things were more 'boring'. Lived experience in this way 'unfolds' in ways informed by speculative practices based on varying degrees of formal epistemological basis. These affective aspects of expectation and anticipation also shape politics and experiences of political manoeuvring space:

> The present is governed, at almost every scale, as if the future is what matters most. Anticipatory modes enable the production of possible futures that are *lived and felt* as inevitable in the present, rendering hope and fear as important political vectors.[7]

Scheduled calamities on the horizon, whether they are already in motion (as in Aponte), or are simply expected (e.g. a major volcanic eruption waiting to happen), produce 'a sense of looming time limits that generate urgency and anxiety about acting now to protect the future'.[8] Once uncovered, the disaster risk may not yet a full-blown emergency, but it is rendered real in its consequences on the present given that knowledge of a threat will, at least in theory, generate responses, both affective and those related to mitigation and emergency preparation.

At the same time, this process also entails a forced passage through affect, in the sense that the anticipatory regime cannot generate its outcomes without arousing a 'sense' of the simultaneous uncertainty *and* inevitability of the future, usually manifest as entanglements of fear and hope.

Life in the shadow of calamity is in this way not only defined by dread produced by the awareness of the hard times ahead; uncertainty as to when things will take a turn for the worse or how things will play out are also central to this affective state. Although responses are made based on the best available knowledge, this knowledge itself remains uncertain. This is made evident to observers because the state of the art on the issue is continuously adjusted and hinges on continuous assessment and re-assessment. Communities where a significant disaster risk is detected hence become subjected to an entire set of measures and formal processes, initiated in the name of safety, but that also end up defining not only the local identity but also the experience of living with these regimes of disaster risk management. Such interventions are often experienced as oppressive, rendering the communities discovered to be at risk as voiceless, as they are subjected to technocratic monitoring and a rendering technical of their own experience.[9] Moreover, living under such regimes also affects mental health and social functioning:

> The effects of living under anticipatory modes of engagement affect physical, mental and emotional well-being in ways that are beginning to be understood as long-term chronic disorders and stress-related disabilities.[10]

This distress arises from finding oneself, as an individual, household and community, in excessively long periods of perpetual alertness— preparing and waiting for the inevitable, yet uncertain arrival of a scheduled disaster (or disaster peak in the case of a slowly emerging disaster)—and as such unable to plan one's life, future or otherwise achieve any sense of predictability or personal/communal security. Living in anticipation of expected disasters can shake one's ontological security for a prolonged period of time, owing to this feeling of remaining 'ready for, being poised awaiting the predicted inevitable' that, in turn, 'keeps one in a perpetual ethicized state of imperfect knowing' as new information is discovered, events fail to come to pass, and as the situation becomes more protracted than previously imagined.[11] Unmet promises and guarantees, in terms of both assistance and the failure of the scheduled disaster to arrive 'on time', reveal to the community the real limits of expert knowledge and its power to predict. This, in turn, intensifies the need for better methods of prediction while also reminding at-risk populations of the uncertainties they are forced to endure. Multiple evacuations will often take place as a result of false alarms, exhausting communities living under such regimes. Some may wish to move and settle elsewhere, but the absence of appropriate compensation and the difficulties associated with selling property areas publicly defined as high-risk can make it hard to move on. In other words, uncovering danger on the horizon 'authorizes pre-emptive actions in the present' that are 'forced by a purported urgency in the future, legitimating, destroying, removing and/or eradicating now' in the name of disaster risk.[12] In this sense, 'life in the present is held hostage' to a disaster not yet real in its material destruction, but real in its effect on everyday living.[13] We may infer that disasters which manifest themselves slowly can produce similar feelings of dread and anticipation, as communities wait for their scheduled peaks, and their eventual climax.

One could argue that the study of how the lived experience of dread and anticipation shapes the present will be increasingly important in understanding a host of risks we face across levels, ranging from antibiotic resistance or climate change at the global level, to anticipated massive earthquakes at the local level. In their 2019 piece *Experiencing Anticipation*, anthropologists Christopher Stephan and Devin Flaherty set out to ground the notion of anticipation within the broader scholarship on socio-cultural views on temporality.[14] Arguing that the increasing future orientation of society warrants greater attention to analytical approaches to how humans construct present-future cultural understandings, the authors recognize the inherent value in studying events as they unfold and as they are being dreaded.

Citing Vincanne Adams and colleagues, the piece notes that many connotations of anticipation are often implied simultaneously in discussions of the concept.[15] Not all of these take a phenomenological view focusing primarily on lived experience. Some conceptualizations of the term have focused on the way in which speculation or information concerning predictions or forecasts impact actions in the present. These framings may be contrasted with other branches of scholarship that imply less formal ways of knowing, such as more-or-less informed ways of imagining potential futures. Some of these works are identified as based on the perspective of lived experience or conceptualizing anticipation by way of affect or anxiety. More applied research has also looked more specifically at what kinds of actions are triggered in the present and how those futures have been approached in the present in a more practical sense. Lastly, the authors identify a body of work that emphasizes the technologies of anticipation more explicitly, such as work centred on forecast methodologies, techniques for pre-emptive action and other technicalities related to anticipatory action. As mentioned, it is not always clear which one of these conceptualizations are implied across the wider scholarship of anticipation, and some publications may employ several at once or implicitly concern both lived experience perspectives and technical aspects in the same contribution, without explicit mention.

The experience of anticipation and its associated emotional manifestations can be understood as intersecting the natural and the social. The environment has certain features that may be hazardous to humans (these dangers may also be intensified by humans through such practices as deforestation, which increases the risk of many types of geological hazards), and the social environment shapes how these hazards are experienced and the impacts they have on society. An experiential perspective on anticipation thus attends to both the social structures that shape lived experience as well as the more objective conditions by which lived experience is ultimately constrained. In other words, life in anticipation is essentially about studying how time and temporality are experienced when shaped by the presence of a sense of imminence (of a type of adversity). An experiential view is therefore useful for grasping connections between 'the world as simply being "out there" for actors to encounter' and the way that 'world-making' takes place within socio-cultural dynamics.[16] Engaging anticipation as an analytical construct does not imply that the phenomenon is coherent and free of contradictions, however. As Christopher Stephan and Devin Flaherty argue:

> A key component for studying anticipation from this perspective is recognizing and tracking how the future manifests across the range of practical and reflective engagements in everyday situations. As anthropological and phenomenological theorists have long argued, the first-person perspective is in large part lived from within a pragmatic and unreflective engagement with everyday activities. As such, embodied competencies, perceptual habits and a naturalization of our social and material world provide an assumptive background against which anticipatory experience takes shape. Yet people also shift into more reflective, critical or sceptical perspectives in which anticipatory experience may figure as a significant part of re-examining, reconstituting or recommitting to the social world. It is often 'breakdowns' or disruptions to everyday expectations that catalyse explicit acts of reflection (in this case, on the future).

We thus see that attending to the analytical notion of how life is shaped by imminent or expected adversities also implies remaining attentive to the elusive and oftentimes incoherent nature of lived experience. Therefore, in theorizing life in anticipation, a central challenge is to strike a balance between narrative flow and analytical nuance. As all scholarship to some extent hinges on imposing some kind of narrative order on messy data (accounts of the lived experience of an individual may not even seem entirely graspable even for the person in question), the importance of reflecting on this dilemma should not be underestimated. Furthermore, observations and narratives on affect, lived experience and temporally protracted risk experiences are made sense of also through cultural filters, in addition to cognitive and sensible ones, among others. The implications of this interplay of sense making and meaning-making on lived experience is that, over time, personal experiences and accounts of events are shaped by day-to-day and week-to-week variations in degree of worry, the occurrence of secondary stressors (a divorce, death in the family, crop losses, or simply a change in discourses in the news media etc.) or other factors that render the unfolding of temporal experience ridden with inconsistencies. In hindsight, however, many of these nuances may well be combed over as our brains seek to establish narrative coherence.

Anticipation as a mode of experience is ultimately shaped by attention or acts of noticing, which wax and wane over time as the issue salience increases and decreases as a result of internal cognitive processes and emotional digestion, in combination with external stimuli, either social or directly connected to changes in the disaster risk picture itself. Along these lines it has been argued that 'cultural configurations of thought, affect and perception' are ultimately focused through ways of being attentive, and hence 'play a fundamental part in shaping and lending relative salience to anticipatory experiences'.[17] Yet the ways in which these acts of noticing are experienced ultimately depend on personality, circumstance and culture, rendering the impact of feelings of anticipation on lived experience relatively fluid:

Founded in the vicissitudes of attention to the future, the content of anticipation can appear in a spectrum of precision ranging from vague and impressionistic to articulable and precise…; it is therefore problematic to presume that anticipation is always experienced as a clear and distinctive previewing of future potentialities. Anticipatory experience flux in time. What might appear from the outside or in aggregate as a relatively stable vision of 'the future' is always to some extent an abstraction from lived experience, within which even anticipations of 'the same' future can vary greatly in their particular content and scope of their anticipatory horizons disclose not only present-day concerns, but their felt thrust, which may at times lie just beyond the cusp of articulability.

The experience of anticipation is at once both shared and individual. Just as different kinds of people respond differently to risks, people respond differently to actual adversities. Temporal experience may in this way be understood as culturally maintained while also being individually amplified or lessened. Indeed, the lived experience of the flow of time (as with the advancing of disaster onsets) is, to a large extent, shaped by the uncertainty of what the future brings and thus requires a sense of planning, in a soft sense of the word. From a phenomenological stance, then, we can distinguish 'clock time' from 'lived time' (or 'social time'),[18] the experience of time's passing being an example of the latter and specific points in time according to a time standard as an example of the former (e.g. an onset of five years, two hundred and sixty weeks, contrasted with the temporal experience of living in suspense or dread for what feels like a seemingly endless period of time). However, the distinction between these two forms of experiencing time should not be understood as mutually exclusive.

The object of anticipation will tend to not remain static in time (due in part to the uncertainties of prediction mentioned earlier), and the scheduled disaster will tend to be rescheduled time and time again until the expected adversities eventually materialize fully. Even on occasions where time is experienced as a continuous unfolding of everyday living, specific anticipated events could well occur and indeed come to pass, with a sense of lingering anticipation. Hence, 'anticipatory temporalities will not necessarily correspond with a linear progression of time and succession of events',[19] but will instead be characterized by elusive endpoints and aftermaths more akin to slow, gradual fade-outs than to abrupt returns to normalcy (following a period of 'recovery'). However, as the authors remind us, our affective modes will ultimately also be affected by interaction with others, ultimately producing a co-construction of reality particularly informed by technical data concerning the natural world (i.e. the geological hazard in the case of Aponte) in combination with narratives based on people's meaning-making. At the individual level, then, personal interpretations of one's own experience are possible, but oftentimes severely constrained by the daily interactions that ultimately also serve to socialize not only our identities, but also our options and emotional state. In other words, our temporal experience will ultimately be shaped by 'the lived time of others, both at the level of concrete face-to-face relationships and within broader communities of practice',[20] although not perfectly calibrated to these.

Central to Christopher Stephan and Devin Flaherty's discussion of the analytical notion of anticipation is the concept of 'anticipatory object'.[21] The object of anticipation may be any kind of expected event or process already-in-motion, which is anchored temporally in some way. That is, the scheduled arrival of an expected calamitous outcome is usually somehow 'already determinate and figured in time',[22] albeit elusively. In this sense we can more fruitfully understand anticipatory objects not as static scheduled events but as expected adversities whose advent lingers on the horizon, but rarely arrives on time; their onset dynamic more often than not fail to progress exactly as modelled or as anticipated.

In the context of the Aponte case explored in this book, we may say that the anticipatory objects are twofold, with one operating on the hazard side of the equation and one on the disaster side. In reference to the hazard process, the anticipatory object is the 'aftermath' or culmination of the geological movement, when the hazard has reached its peak intensity and/or the hillside dislocates, causing a collapse. However, we also have a social aspect to this process where present-day lived experience is also very much shaped by anticipating the consequences of the hazard—its material destruction and the eventual uprooting and displacement. Most of these emotions are best described as melancholic, anxiety-inducing, trepidatious and other strong experiences of distress and worry. At the same time, given that after a period of mourning the community is accepting the severity of situation, the eventual resettlement and reconstruction process is also looked forward to in a somewhat positive sense. The aftermath here thus suggests two outcomes that may occur simultaneously: the total destruction of the worst-affected parts of Aponte, as well as the construction of a new Aponte, likely nearby the surviving parts of old Aponte. Both of these outcomes conform to the temporal anchoring described by Christopher Stephan and Devin Flaherty, in the sense that the aftermath is continuously rescheduled further and further into the future as the crisis becomes more and more protracted. At the time of writing, neither the dislocation nor the resettlement have materialized, but both continue to be anticipated. Thus, expectation and anticipation may, on the one hand, be aimed towards an outcome whose arrival is dreaded, but it may also be thought of as a potential source of frustration when things do not come to pass, further exacerbating feelings of uncertainty:

> How any experience of anticipation will correspond to a realized or unrealized future will depend not only upon the actual course of events, but upon the collective and personal conditions within which the anticipatory experience is founded'.[23]

In a 2017 essay on *The Politics of Anticipation: On knowing and governing environmental futures*, Céline Granjou, Jeremy Walker and Juan Francisco Salazar approach anticipation from the viewpoint of futures studies. Serving as an introduction to a special issue on the topic, the text sets out to draw attention to social aspects of the notion that 'the future could always have been otherwise'.[24] Tracing the politics of anticipation back to the realization that views on the future inherently concern ways to exist on our planet, at any scale, thus serves to unpack the implications of engaging with the concept of anticipation in the present. The political aspect of anticipation is made evident as thoughts on 'the future' ultimately 'inform action in the present' in the form of 'ways of knowing, forecasting, and actively anticipating future events' which, in turn, constitute 'crucial elements of social organisation'.[25]

Indeed, anticipatory politics takes place across scales and is familiar to everyone as a timeless aspect of human existence, albeit it manifests in different form in contemporary times due to improvements in monitoring and forecasting technologies. Global environmental mega-trends such as biodiversity loss, climatic change and the spread of infectious disease are closely monitored in current times, giving rise to previously unavailable forms of experience. These forms of detection enable us to perceive phenomena, so to speak, before they become unambiguously noticeable to the human senses, affecting how they are anticipated and the ways in which they create new forms of dread by engaging alternatives to direct experience in the form of visual technologies like illustrations, documentary films and statistics. Nevertheless, anticipation also operates on a local or even personal scale, as exemplified by expectations for harvests, local risks or personal tragedies. The selection of which menacing risks are attended to and which are downplayed is in itself inherently political. A central implication arising from this realization is that scholars of anticipation ought to critically examine 'the ways in which emergent threatening futures are [made] known, anticipated, fostered, pre-empted, and prepared for'.[26] What characterizes life in anticipation, from this point of departure, is that an anticipatory orientation produces and indeed demands action in the present so as to shape the nature of that which is expected. Crucially, attention to the political aspects of life in anticipation must not ignore the tendency for many speculative predictions of future potential occurrences to also be disregarded or ignored. In other words, life in anticipation can be thought of as also being about 'enacting a future that (hopefully) makes a present that (hopefully) shapes the future'.[27]

Detecting a considerable disaster risk in a place is only a first step towards proper management of that disaster potential. Exercising politics on future potentialities is fraught and often becomes controversial as it can be experienced as inconvenient in the present for affected communities and disaster managers alike. The same can be said when surveys discover that a slow-onset phenomenon is already in motion, giving rise to the realization that symptoms observed in the present are mere precursors to the calamity that may materialize once the onset has advanced further. Drawing on the work of geographer Ben Anderson, Céline Granjou and colleagues observe that acting on risk knowledge implies a 'translation of foresight into preparation, precaution and prevention', although this sequence of events 'cannot be presumed'.[28] After all, as the authors remind us, 'there are times when time's up: when it is no longer possible to shape and transform our lifeworlds in ways that we might prefer'.[29] Slow onsets should for this reason not be understood at affording infinite decision time.

In his 2010 article entitled *Preemption, Precaution, Preparedness: Anticipatory action and future geographies*, Ben Anderson examines the connections between anticipation and response through the concepts of practices and logics. Serving to unpack the ways in which expected futures impact politics (and justifications for action) in the present, the author observes that anticipatory responses are often shaped by 'a seemingly paradoxical process whereby the future becomes cause and justification for some form of action in the here and now'.[30] In turn, untangling this paradox requires us to better grasp the ontological status of future calamities in the present; or how and in what sense they are real before they have even happened. After all, each 'attempt to stop or mitigate a threat holds certain assumptions about "the future"' and potential disasters it could bring, although it is equally clear that 'the future will radically differ from the here and now (even as the here and now or the past may contain traces of the disaster yet to come)'.[31] We know that the future is contingent, in the sense that perfect prediction is unavailable to us.

The future is the realm of risk. We may find clues and signals, but not foresight. It tends to 'exceed present knowledge' as well as 'the capability to generate knowledge'.[32] Its nature is potentially quite different from the past (history can only teach us so much); as the future draws nearer, the potential for a contingency to arise is always present—although only some calamities appear unforeseeable in hindsight. In effect, our preoccupations in the present are generally directed towards surveying the horizon for potential adverse futures we seek to prevent from becoming tomorrow's reality. The 'presence of the future' is rendered real through 'epistemic objects' like models, projections, statistical trends, images, reports, or more affective states such as rumours, fears, anxieties, dread and hope, each triggering its own set of potential responses with the purpose of shaping outcomes.[33]

Ben Anderson distinguishes between practices and logics in his conceptualization of anticipation and actions it triggers in the present. These practices of calculation, imagination and performance we can understand as formalized anticipatory practices. Calculated futures are, in essence, about the way in which models, data, predictions and other calculative methods, as well as trends (often assumed to follow a pattern similar to the past when they are projected into the future) may steer action in the present. Secondly, their contingencies are dealt with not so much through calculation as imagination. We recognize the importance of qualitative approaches to anticipation in the form of scenario building, inference from experience, envisioning potential futures or even those risks that have never been experienced but remain relatable for us through works of fiction (or conceivable 'black swans' we attempt to imagine without prior experience). Thirdly, some futures are performed rather than imagined or calculated. In essence, linked to imagining, they also contain an element of embodiment in the form of role-plays, acting or exercises. Ultimately, these kinds of anticipation not only allow for theoretical knowledge of some potential future event, but also open up for simulated experience and, indeed, a sense of theoretical yet practical readiness.

Part of living in anticipation of some future adversity also involves efforts at cushioning these. Ben Anderson refers to this as 'logics', where he identifies precaution, pre-emption and preparedness as distinct types. These three logics relate to the future in distinct ways. Precautionary actions essentially set out to foreclose some envisioned or feared future adversity by stopping action in the present on the grounds of future risk. Examples include stopping a scientific experiment or cancelling a large infrastructure project on the grounds of a suspected future adversity. Common in these situations is that processes in the present are either terminated or avoided on the grounds that the risks do not justify potential advantages.

Pre-emptive action, on the other hand, is rather about acting in the present to prevent a feared outcome that may not be stopped simply by acting or *not proceeding further*. Pre-emptive action is reserved for that category of contingencies where the process itself is not a result of the agency of the agent carrying out the anticipatory action (e.g. pre-emptive strikes are done by an agent to impact the agency of another agent, whereas precaution would involve simply not striking out of fear for what might happen as a result). Preparedness actions are reserved for those contingencies that cannot be altogether mitigated or where a residual risk would remain regardless. Preparedness relates to the future by maintaining a sense of readiness for that which cannot be feasibly shaped in the present through precaution or pre-emptive action.

As the Aponte case demonstrates, these types of practices and logics are not, in a typological sense, mutually exclusive and exhaustive. The Aponte disaster is subject to both technical calculation and social imagination and is also performed through expert visits, traditional rituals as well as the usual dynamics of disaster politics. In terms of logics, the way in which the Inga live in anticipation of the potential escalation of the disaster necessitates preparation and a maintenance of constant readiness for continued destruction and risk of displacement. At the same time, the planned relocation process can be considered a form of pre-emptive action aimed at getting people out of harm's way (hopefully) before the hillside collapses, converging the current slow-onset phenomenon into a rapidly occurring one.

An alternative way to understand the temporal aspects of lived experience is through narrative methodologies, as in sociologist Ruth Fincher's 2014 article *Time Stories: Making sense of futures in anticipation of sea-level rise*.[34] Her work chiefly aims to study how people respond and adapt to certain changes in their environment, both short and long term, particularly in response to signals about expected future changes. The narrative approach is in this way a methodology for exploring 'how people's lived experiences of time', particularly local changes observed by a person or household over a lifetime, 'are related to the ways they respond to information about a distant future'.[35] Time stories are conceptualized as narratives concerning envisioned futures in reference to the present and the past. Narratives are what binds past, present and future together into a relatively coherent experience, although such stories and memories of temporal experience are rarely coherent when multiple story fragments and pieces of experience have to be synthesized into larger stories about a place or a household. Narrative approaches to life in anticipation will, in this way, also have to account for the complexity and incoherence of experience that occurs as a result of imperfect memory and the real difficulty of placing large chains of events in the correct order. People draw on these stories about temporal experiences to make sense of the trajectory of time's passing with reference to their own lives and changes in the observable environment around them.

Ruth Fincher's conceptualization of temporal experience distinguishes between the concepts of 'time stories' and 'time practices' that operate at four different geographical perspectives on time.[36] Time stories are those stories people construct and remember (remembering over time and telling stories in the present are distinct mental processes) about their narrative self, the experiential present and their imagined futures. In other words:

> [T]ime stories fix time in places and places in time. They include (often explicitly) assessments of what is just and fair and should happen in the future: the end points of time stories are often preferred futures, or suggestions about how presents or futures can right pasts.[37]

Time practices, on the other hand, are actions directed at correcting past injustice, shaping the present or ways of engaging with future anticipation, such as investment, savings or adaptation measures. These stories and practices operate at different scales.

One level at which temporalities operate is in the way it is felt and understood materially. Places and people's experience of them change over time. Some of these changes are measurable in the objective sense, such as deterioration or improvements in ecosystem health or changes in local mean temperatures over a long span of time. Observations also shape experience, but they are more prone to the inconsistencies of memory and problems of narrative coherence; remembering how things were is not easy. These temporalities may range from daily changes to seasonal, all the way up to the glacial or geological timescales.

A second level of experiencing temporality is through mundane everyday tasks such as waiting, or the experience of time as it is 'lived', so to speak. Examples include plans, decisions to be spontaneous (to surprise, seem untimely), temporal structures, such as institutionalized routines (school, worship, meetings), as well as functional aspects (coordinating to meet at a certain time). This daily life or life as it is lived approach to temporal experience directly intersects with the material one because the passing of time is, in one sense, material, but in another sense purely experiential in a cognitive or social aspect. We use material signals of time's passing to orient ourselves temporally (e.g. the tide, moon phases, changes in the environment, seasons, day and night), but even if one were deprived of all senses one would still experience the passing of time as an experience within one's own head.

A third way of engaging with temporal lived experience is through thoughts about and memories of pasts, presents and futures. Citing the work of Elizabeth Grosz, the third kind of experience of time is experienced as a form of 'extended present' and is described as follows:

> The present itself, always a continuous present that never passes into the past, is nevertheless not present to itself. It remains fractured and refracted through reminiscence and anticipation, the murmurs of the past and the potential of the future. Every present is driven by memory … equally, the present always spreads itself out to the imminent future, that future a moment ahead for which the present prepares itself by reactivating the past in its most immediate and active forms, as habit, recognition, understanding.[38]

A fourth way of experiencing the future in the now is in terms of future-oriented or anticipatory practices in the present. Narratives about futures, both gloomy and optimistic, shape life in the present by means of anticipation. Narratives surrounding the changing nature of crises or that the disasters of tomorrow will not be like those of the past are just two examples of future-oriented narratives with practical implications in the present. Considerable action is also taken to curb projections and worries about potential future realities, such as aims at flattening the curve during pandemics or shorting a collapsing market. We see that these practices often emerge out of representations, such as representations of how the future of development, energy, health or technology will turn out. While these projections never quite seem to materialize, we also know that the future is never quite like the past or the present and that success in life hinges, to some extent, on assuming productive practices as part of its anticipation. This is nothing new and we have seen this for a very long time (e.g. Malthus envisioning population bombs, the end of poverty, the fall of the capitalist system or a dreaded cataclysmic or existential risk). In this way, 'the everyday is the temporal site at which events and meanings at different temporal scales coalesce for people making sense of their situations', mainly because time stories are the way in which 'people's material experience of time in the past and the present are actively associated with their thinking of what the future will be like'.[39] Of particular relevance to the Aponte case, drawing on Ruth Fincher's time stories concept as applied in the context of climate change, is that climate change is an example of a temporally multilayered phenomenon in the sense that it forces us to think about tipping points and potential regime shifts. Returning to Ben Anderson, a precautionary logic suggests that the ideas of distant potentialities must at the very least be entertained and that linearity of onset cannot be assumed a priori, whereas political realities and lived experience often focus on more pragmatic and immediate concerns with present-day (material) implications.

Dealing specifically with the topic of temporal perspectives on the experience of displacement we find Bahar Sakizlioğlu's 2013 article, *Inserting Temporality into the Analysis of Displacement: Living under the threat of displacement*. This ethnographic study theorizes the experience of displacement following the reception of news of gentrification and demolition of neighbourhoods in a Turkish setting. The study engages with the question, 'How is it to live in a house that is to be demolished?', among others.[40] Central to our inquiry into living in anticipation of disaster and displacement, is the article's focus on the affective state that emerges after a community receives news of impending displacement. What often starts out as rumours or a lingering feeling that displacement may be an option, before being formalized politically in plans, renders the displacement process affectively real before it occurs materially. Citing Bourdieu, she goes on to argue that anticipation, or waiting, is in itself a key characteristic of governmental power over such bureaucratic processes:

> Waiting is one of the privileged ways of experiencing the effect of power, and the link between time and power – and one would need to catalogue, and analyze, all the behaviors associated with the exercise of power over other people's time both on the side of the powerful (adjourning, deferring, delaying, raising false hopes or conversely rushing, taking by surprise) and on the side of the 'patient' as they say in the medical universe, one of the sites par excellence of anxious powerless waiting. Waiting implies submission.[41]

In effect, life in the face of displacement as recorded in this study suggests that, on the one hand, political authority is used to govern the temporality of such projects (which are frequently continuously delayed and postponed), generating intense insecurity and inability to plan on the part of affected communities. While political agency is eroded, the social aspects of community also slowly collapse as people move away or enter into a state of despair.

Oftentimes impending disasters are known to be a risk by affected communities, but processes to mitigate risk are delayed by bureaucratic temporalities, such as in the case described by Karine Gagné in her 2019 article *Waiting for the Flood: Technocratic time and impending disaster in the Himalayas.*[42] The paper concerns the Phunktal, who lived in the shadow of an anticipated flood due to a natural dam that had accumulated upstream. The dam was expected to burst after a certain amount of time, but little was done to mitigate the looming disaster.

One key aspect of how this book sets out to theorize life in anticipation of risk and disasters is the issue of bureaucratic delay and inaction in the face of expected disasters, often occurring in communities with a history of neglect by the state and the ruling elite. The management of disaster risk and anticipation of disasters is, as we know, inherently political. Although it is in some cases also a result of scientific limitations and lack of prior knowledge of hazardousness, even forewarned disasters are all too often allowed to occur. One prominent Colombian example is the 1984 Nevado del Ruiz disaster, also known as the Armero disaster, which triggered a politics of anticipation in hindsight as part of the accountability process.[43] Ultimately, spaces subject to anticipated risk become subject to 'regimes of anticipation', which, in turn, 'increases the control that central authorities exercise on local communities in the name of expertise and safety'.[44] In this way disaster risk reduction can, in some instances, be experienced as a deliberate rendering technical of mitigation efforts in the form of expert vocabularies and technological warning and monitoring systems. In effect, life becomes reorganized in response to expected disasters, which constitutes what elsewhere has referred to as 'the productive nature of anticipation'.[45] As Vivian Choi notes, this power discrepancy is not always unproblematic, as 'state attempts to control the future [may] remain in constant tension with the attitudes and opinions of people'.[46] As Karine Garné explores through the case of the Zanskarpas, this anticipatory regime led the local community to feel deeply insecure and distrustful of the intentions of central authorities, as the disaster was known to be imminent, but nothing appeared to be happening in terms of response measures. The neglect of authorities was thus experienced as a form of oppression that forced people into a underprivileged status of waiting, coming back to Bourdieu's points about waiting and power. To take a local perspective:

> in the minds of many people in Zanskar, the state's interventions were 'just a show' that involved unnecessary mobilization of outside experts while ignoring local knowledge. This was a play for which the script was already written: Zanskarpas knew the disaster was going to happen [yet little was done].[47]

Therefore, lack of clear communication, transparency and accountability can arguably feed into feelings of marginalization and thus exacerbate the dread that finding out about one's own disaster risk exposure may produce. This affective sate leaves households feeling that they are being governed by technocratic forms of anticipation that hardly seem accessible. Representatives of anticipatory regimes often speak their own language (indeed, disaster risk reduction terminology is frequently criticized for being inaccessible and rather full of jargon,)[48] to planning processes that seemingly live their own lives, as well as temporal logics that do not follow the temporalities of hazards to which communities are exposed. In situations where there is little communities can do for themselves (in most cases they will, if the hazard or disaster risk can in some way be augmented through local action), this sense of marginalization creates a deep sense of despair.

Writing on the experience of living with earthquake risk in the San Francisco Bay Area, Charlotte Mazel-Cabasse, in her 2019 book *Waiting for the Big One: Risk, Science, Experience, and Culture in Disaster Preparedness*, explores how projected disaster acts on the present.[49] Noting that disasters exist in several modes simultaneously, both as memories, expectations and calculated probabilities, she centres on the epistemological relationship between technoscientific understandings of likelihood and the experience of living in the face of inevitable future disaster. While each individual experiences the risk distinctly, she also notes that the experience itself gives rise to a shared project that arguably also shapes the culture in a place:

> Waiting for The Big One is also a collective experience that brings together both experts and non-experts, who collaborate to mobilize residents' attention and concern. Looking at the many different ways an earthquake exists—not only as geological process, but also as a past and potential future threat—help unfolding the everyday experiences of concerned individuals in the San Francisco Bay Area, and illustrate how a preconceived framework of analysis often fails to recognize the complexity of 'living with risk'.[50]

Disaster scholars increasingly find that the temporal aspect of hazard and disaster experience requires better conceptualizations of disaster ontologies, and this book has set out to do just that. A better understanding of how disasters that are known to occur in the future (but have not yet happened) as well as how disasters that are currently happening, albeit slowly (and seemingly unstoppably), cannot be covered adequately by the hazards paradigm and must therefore be brought into the disaster paradigms as part of scholarship on the lived experience of such phenomena. While everyday life is indeed full of distractions that steal our attention away from what is, in reality, known to be impending doom for many (such as in the San Francisco Bay Area under a 'Big One' scenario), distraction is only possible to some extent. It ultimately gives rise to cultural manifestations of living in the face of imminent adversity, knowing that things will eventually become very ugly, even if the exact timing remains unknown.

Although the earthquake itself is a hazard and the theoretical possibility of avoiding disaster is always there, few would dare to argue that disaster is not preconfigured in such places. Such a 'missing link between the known and the unknown' vividly displays the 'limits of science as the only resource which to face the threat', as people ultimately also rely on such notions as luck or fate, or even belonging, perhaps as mechanisms of coping with life in places shaped by the threat of disaster.[51] To summarize, the ontological security of people living in places exposed to both slowly emerging disasters and scheduled disasters is continuously threatened by the knowledge of this fact. Yet, coping mechanisms, as displayed through either practices or narratives, seek to maintain a sense of existential security in a place, retrofitting or preparing being an example of the former and discourses emphasising the collective experience of disaster being an example of the latter.

7.2 Ontological security and living through slow calamity

As our knowledge of the hazards around us increase and disasters become visible before they even occur in a material sense, our fear of disaster risk also increases. Susceptibility to disaster is not only felt individually, but is also felt socially as 'environments of risks'[52] subjecting large groups of people in risk regimes. Sometimes these experiences are universally lived across all areas of the world (as in the example of global phenomena such as climate change or COVID-19), and sometimes they are experienced more locally. Common to disaster experiences is that they involve a degree of shattering of trust in the order of things, a condition shaped by feeling that ordinary life is disrupted (either suddenly or gradually) by forces that seem overwhelming and threatening. In other words, menacing disasters threaten our lived experience of ontological security.

Living in anticipation of unfolding or expected disasters demands an exceptional level of trust in technical expertise and expert knowledge on the part of affected communities. Taking such leaps of faith and trusting that monitoring and planning efforts will prove sufficient is often mentally costly. It is not uncommon for communities to express ambivalent attitudes towards such relationships. Although technical knowledge is generally held in high esteem and associated with high levels of respect, laypeople may often find themselves feeling like leaving such high-stakes questions to people who are not at risk themselves is difficult. After all, 'trust is only demanded where there is ignorance',[53] meaning that if we consider leaps of faith to be central to trust, taking a leap of faith in the ability of prediction to save us from disaster risk is something with which many are uncomfortable. Living under regimes of disaster risk put in place to cushion against slowly developing dangers or expected future disasters thus necessitates trust in those systems. However, trust in those systems is also normally ambivalent.

Ontological security, as conceptualized by the sociologist Anthony Giddens, has been described in terms of a 'protective cocoon'.[54] Our lived experience of the continuity of things and trust in that things will remain, at least to some extent, predictable, is for most people essential to their feeling of existential security. Except perhaps for some individuals who have little need for predictability and stability to thrive, ordinary lifestyles are typically based on a sense of predictability, even in situations perceived as outsiders as precarious. The power of habit is strong and what looks like a terrible life to be leading by some observers may be normal to the person observed. Ontological security hence refers to a very wide form of safety and trust in an experiential sense. Despite hardships and trying times which cause our plans to fall short in the face of unforeseen personal and societal contingencies, self-identity remains somewhat intact in most cases. Disaster and calamity has the power to shatter this sense of continuity and predictability, at least temporarily. Populations affected by earthquakes have been reported to struggle to sleep indoors for some time after their traumatic experience. The violent shaking and structural damage that earthquake hazards trigger shake ontological security to its core. Seemingly everything surrounding one may turn to dust and rubble in seconds. How is one to 'trust' that every moment of lived experience thereafter will not be suddenly and unexpectedly interrupted by violent shaking, death and destruction? This sense of basic trust in that things 'generally but not always' stay the same is the essence of the protective cocoon and its emphasis on trust in the continuity of the present:

> Ontological security is one form, but a very important form, of feelings of security in the wide sense in which I have used the term earlier. The phrase refers to the confidence that most human beings have in the continuity of their self-identity and in the constancy of the surrounding social and material environments of action. A sense of the reliability of persons and things, so central to the notion of trust, is basic to feelings of ontological security: hence the two are psychologically closely related ... [it] has to do with 'being' or, in the terms of phenomenology, 'being-in-the-world.'[55] But it is an emotional, rather than a cognitive, phenomenon, and it is rooted in the unconscious.

Giddens notes a distinction between productive forms of fear, such as that fear or dread which is central to the way we understand the experience of living in anticipation of disaster, and more disabling experiences of undirected anxiety. Should the earthquake survivor fail to ever recover a sense of trust in the general stability of things we would be talking about a dysfunctional type of response or a trauma that causes long (perhaps lifelong) pain. The expectation that we should recover from a psychosocial perspective, however, implies that at some point ontological security is expected to have been more or less restored. After countless days of experiencing non-shaking we would again trust in a shake-free tomorrow. Still, we would also be aware, in a new sense, that shaking is possible and indeed unpleasant, giving rise to what some have labelled disaster subcultures.[56] These are, Giddens argues, distinct from undirected senses of supersensitivity to risks and potential adversities. In general, populations live with a pragmatic acceptance of some types of risks and hazards, and the degree to which they enter into and are made real in the everyday depends on the intensity of these risks to which one is exposed. When the routines of everyday life are disrupted by the unexpected dominance of anticipatory risks, they are lifted from their previous status as background noise (to be ignored) to potentially deadly risks to be anticipated (and acted on or intensely feared, depending on, among other things, their degree of perceived influenceability).

A distinction here may be made between the impact that sudden and acute types of disaster have on ontological security and the experience of more elusive kinds. Risks that have more of a 'menacing appearance' or risks that give rise to more of an 'unnerving horizon of dangers' produce a different type of stress than disasters that deliver their destructive potential in one blow.[57] This is not to say that one experience is worse or less traumatic than the other. Slow violence has been argued to be distinct, but equal.[58] As I also argued in Chapters 3 and 4 of this book, more elusive and slow-onset kinds of phenomena first and foremost have to be made sense of in a different way, as their very existence is often obscured, at least initially, by signals that may be difficult to piece together into a coherent whole. In other words, grasping the severity of the situation may be challenging when it is not rendered immediately obvious to spectators. As many of us will be able to relate to more than ever because of our experience of the COVID-19 pandemic, the way elusive phenomena are described in the news perhaps shape the meaning-making process more than our direct observations connected to such phenomena. As Giddens describes, elusive and menacing disasters are in some ways experienced simultaneously as processes that we observe, feel, imagine and, to a great extent, as images and stories presented to us (in a pre-digested fashion) through the news media and official communication. They gradually pierce our protective cocoon and to leave us with a sense of compromised ontological security by intruding and seeping into everyday lived experience through our (slowly) growing awareness of their existence, thus rendering them real (to us). Occasionally, even their reality, so to speak, may be more a product of media discourses than by our observation themselves:

> Many of the events reported on the news, for instance, might be experienced by the individual as external and remote; but many equally enter routinely into everyday activity. Familiarity generated by mediated experience might perhaps quite often produce feelings of 'reality inversion': the real object and event, when encountered, seem to have a less concrete existence than their media representation.

Technoscientific artefacts are also a big part of this puzzling together of meaning-making pieces. We observe this tendency everywhere, ranging from our lived experience of climatic change to our experience of antimicrobial resistance. Without measurements or external knowledge of these phenomena we might never have become aware of their existence as a result of an absence of trend data. However, once rendered real through a mix of scientific longitudinal data, science communication, news media reporting, public engagement and social risk amplification leading to cultural embeddedness, these looming crises could not necessarily be observed by individuals based on experience without measurement. Yet, as we know, this does not preclude narratives of 'future disasters' such as these (and many more, e.g. meteor strikes, artificial intelligence, pandemics or super volcanoes) from entering into lived experience in the present. After all, of all of these scenarios are known to affect mental health adversely——not necessarily as a result of direct harm, but due to sentiments of future anticipation.

7.3 Anticipation and ontological security in Aponte

So how are these ingredients for an analytical framework presented thus far of relevance to the Aponte case? The primary purpose of this chapter was to conceptualize the notion of 'life in anticipation' as an analytical lens of relevance to disaster research and deliver on the aims of the book described in the introduction to the book. As argued at the outset of this chapter, life in anticipation is a potentially interesting approach to study both communities living in the shadow of expected future disasters as well as communities currently facing slow-onset forms of devastation. The general discouragement from engaging the concept of 'slow-onset' disaster—because all disasters are slow-onset when underlying vulnerabilities are taken into account—has made it difficult to theorize the concept of 'onset' in disaster research. As readers will have noticed by now, I have written this book without strictly adhering to conceptual consistency, instead focusing on employing an ordinary rhetoric style in my narrative. I have used concepts like disaster, calamity and adversities interchangeably. I have also varied in my use of concepts to describe the same phenomenon, such as having sometimes interchangeably written slow-onset disaster, looming disaster, menacing disaster, gradually manifesting calamity, and other variations of these concepts, oftentimes to consciously underline that what we call the concept is of less relevance to my project here.

My aim here has mainly been to conceptualize the phenomenological aspects of the sensation of living in the face of calamity and disaster. A secondary aim has been to suggest that 'onset' is not only interesting in the study of natural hazards, it is also potentially interesting for disaster scholars. To summarize this chapter, and to illustrate how my notion of 'life in anticipation' (drawing on a much larger literature on anticipation more generally), which also draws heavily on Anthony Giddens' concept of 'ontological security', the analytical lens conceptualized in this chapter has a number of interesting implications for the Aponte case, including the following points.

First, the Inga had effectively lived in anticipation of the peak of a slow calamity since the phenomenon was first detected and diagnosed in 2015–16. In this sense, this book particularly refers to the experience of living through (or living with) a *longue durée* calamity onset as opposed to living in the face of a scheduled disaster. At the same time, as we saw in Chapter 3, there are aspects of the natural hazard phenomenon itself that also gives rise to the latter kind of dread. Although the hazard and its resulting disaster experience continues to progress slowly over time, the notion that a tipping point may lurk on the horizon opens up the possibility that the slow calamity could transform into a sudden collapse at later stages of its onset. We saw in the preceding conceptual discussions that life in anticipation may further be described as dominated by a future-oriented perspective on lived experience. This is definitely the case in Aponte, as cultural life among the Inga has, since the realization of their predicament, become intensely shaped by the ongoing and perpetual, yet gradual destruction around them. This experience is as much shaped by a sense of 'What is next? What is the next thing that lurks on the horizon?' as it is to questions such as 'How much longer will I remain safe?' or 'What will become of us?', which, in turn, is related to my second and third points.

Second, the protractedness of the resettlement planning process for Aponte has led to an additional sense of living in *anticipation* of resettlement. Resettlement is recognized in the disaster literature as a highly traumatic experience both for individuals and for communities. When its implementation is dragged out and continuously delayed, the horizon for this dreaded outcome is also repeatedly pushed into the future, prolonging the lived experience of anticipation that it produces. In some ways, it gives rise to a sense of living in a permanent disaster.[59] At the same time, it also increases the odds that the resettlement will ultimately fail; as time passes, a growing number of people consider leaving the community, rendering them at risk of internal displacement, and with little news and little involvement in the process, distrust grows. At the time of writing, the ongoing COVID-19 pandemic jeopardizes the plan as a whole as it is unclear how political and economic priorities will shift, giving rise to yet another aspect of anticipation. At worst, a failed resettlement process or hiccups along the way leading to displacement of would-be candidates for resettlement can also lead to secondary disasters in the form of cultural collapse.

Third, and connected to the issue raised above, the Inga also live with an anxiety connected to their future cultural and social survival, not only as a collection of individuals but also as a politically autonomous and cultural community with its own distinct traditions and practices. Not only is part of this survival connected to their geography and the physical design of the built environment, but, additionally, the very real risk of social deterioration under which the Inga live ultimately means that the cultural survival of their ways cannot be taken for granted. If enough key persons are displaced due to the protracted situation or because the resettlement fails, the community could be changed to such a degree that one may potentially end up questioning whether their culture survived or remains intact following the disaster recovery programme.

Fourth, all of these affective states of anticipation ultimately assault people's sense of ontological security at different temporal scales. Feeling secure in the knowledge that one's way of life will remain predictable and intact in some fundamental way is essential to a life without overwhelming levels of dread and anxiety. On one front, ontological security in Aponte is threatened by the slowly unfolding calamity affecting the community. They are also unable to plan or achieve a sense of predictability in their predicament because of the protracted nature of the resettlement programme. In addition, the worry that their traditional ways of life, their livelihoods and their cultural heritage will be lost provides a third layer of existential angst. In other words, the nature of the calamity and the disastrous response to it erodes any feeling of ontological security they might otherwise have felt, despite the adversities they face as a result of the natural hazard phenomenon.

I would like to argue, based on my observations, that the affective state of living in anticipation of potential community and cultural deterioration has a stronger impact on ontological security than the impact of the collapsing buildings itself. The Inga clearly express their enduring courage and resilience in the face of hardship and maintain this as a source of pride that is socialized into the culture. While the loss of homes and material assets are obviously also a significant source of hardship, it is the loss of social ties, belonging and cultural expression that is most commonly cited as the source of their anxieties and melancholy. With no clear solution yet so as to the resettlement status, and with a national recovery from the COVID-19 pandemic also to address, it is likely that the Inga and people in the region will continue to live in a state of anticipation concerning the outcome of this process.

In conclusion, the final outcome of all of the primary and secondary calamities remains open at the time of writing. The natural hazard phenomenon continues its slow onset. The built environment keeps collapsing gradually. Displacement continues to occur gradually. Little by little, hope wanes for a successful recovery. Social stress accumulates in a population exhausted by having stared their fate in the eye for five long years. An uncertain pandemic recovery lingers on top of these issues. Although the people of Aponte, Inga and *campesinos* alike, remain committed to prevail again this time as they have in the past, those reserve energies and motivations which have been drawn on previously are nearing depletion. Only time will tell how their lived state of anticipation of calamity is realized in the form of post-disaster outcomes, even given that the post-phase ever truly arrives.

Notes

1 Adams, Murphy and Clarke (2009: 246).
2 Ibid., p. 247.
3 Ibid.
4 Ibid.
5 Ibid.
6 Coyle (2004: 520).
7 Adams, Murphy and Clarke (2009: 248).
8 Ibid.
9 Gagné (2019); see also: Waddell (1977).
10 Adams, Murphy and Clarke (2009: 251).
11 Ibid., p. 254.
12 Ibid., p. 258.
13 Ibid.
14 Stephan and Flaherty (2019).
15 Ibid.
16 Ibid., p. 4.
17 Ibid.
18 Ibid., p. 5; Schutz (1967), Neal (1997).
19 Ibid., p. 5.
20 Ibid.
21 Ibid., p. 6.
22 Ibid.
23 Ibid., p. 11.
24 Granjou, Walker and Salazar (2017: 5).
25 Ibid.
26 Ibid., p. 7.
27 Wilkie and Michael (2009: 504), as cited in Granjou, Walker and Salazar (2017: 7).
28 Granjou, Walker and Salazar (2017: 10).
29 Ibid.
30 Anderson (2010: 778).
31 Ibid., p. 780.
32 Ibid.
33 Ibid., p. 783.
34 Fincher, Barnett, Graham and Hurlimann (2014).
35 Ibid., p. 201.
36 Ibid.
37 Ibid., p. 202.
38 Grosz (2004), as cited in Fincher, Barnett, Graham and Hurlimann (2014: 203).

39 Ibid., p. 203.
40 Sakizlioğlu (2014: 206).
41 Bourdieu (2000).
42 Gagné (2019).
43 See for example: Zeiderman (2016).
44 Anderson (2010), Gagné (2019).
45 Choi (2015: 289), as cited in Gagné (2019: 844).
46 Choi (2015: 302), as cited in Gagné (2019: 844).
47 Gagné (2019: 846).
48 A whole debate has emerged on developments, confusions and disagreements on jargon and terminology in the field. See among others: Chmutina, Sadler, von Meding and Abukhalaf (2021); Chmutina and von Meding (2019), Staupe-Delgado (2019b), Kelman (2018).
49 Mazel-Cabasse (2019).
50 Ibid., p. 37.
51 Ibid., p. 149-
52 Giddens (1990: 35).
53 Ibid., p. 59.
54 Giddens (1991: 40).
55 The philosophers Kierkegaard, Heidegger and Husserl are considered instrumental to how we understand phenomenology and lived experience perspectives.
56 Without citing a particular publication, the Delaware disaster research center has written extensively on the subject. See also: Granot (1996).
57 Giddens (1990: 124–5).
58 See: Nixon (2011).
59 See for example: van Voorst, Wisner, Hellman and Nooteboom (2015), Wisner and Gaillard (2009).

8 Concluding reflections

As readers will by now have realised, the aim of this book is threefold. In writing this text, my focus has been to further our conceptual understanding of disasters by drawing on the case of Aponte in the Colombian Andes. I had hoped to see their situation resolved by the time this book went into print, but I am afraid that as of yet I cannot provide closure to the story presented in this book. Such is the nature of slow calamities; they tend to last longer than expected with recovery remaining an elusive future goal, seemingly perpetually delayed. In fact, if the literature on disaster recovery teaches us one thing, it is that disaster recovery is in itself a fuzzy concept. After all, different households recover at different rates, and we can even question if recovery happens at all, at least for everyone affected. This is not unique for slow calamities, although the slowness of the destruction renders the lived experience of these phenomena distinct during their onset period.

A second aim of this book was to reflect further on the concept of slow calamity and what implications slow forms of destruction have for how we normally think about the concept of disaster. The book focuses, in particular, on lived experience and sense making. A goal of mine, thus, was for the book to contribute towards conceptual advancement in the field. Although readers may not agree with my thinking, my view of disasters or my conclusions, I would still be happy to see the book spark interesting debates in the field.

A third aim of this book is to reflect on (and perhaps carefully introduce) the analytical notion of 'life in anticipation' as an interesting way to think about disasters. The way I envision this notion it can fruitfully be applied both to slow calamities, such as the one presented in this book, as well as to what I call scheduled disasters—places exposed to significant under-mitigated disaster risk expected to one day produce a disaster. In both instances, the real existence of disaster in the present by knowledge either of its on-going onset or its lingering possibility shapes lived experience.

In this account of the Aponte disaster I have emphasised the onset dynamics of the calamity that affected the community. The community noticed the appearance of cracks and small fissures in the ground and in homes in 2014 and early 2015, connected to previous and recurrent mass movements. By 2016, it was clear that the slow geological phenomenon affecting the area would eventually produce yet another calamitous outcome for the community, this time in slow motion. Faced with the very real possibility of community disintegration due to displacement

DOI: 10.4324/9780429288135-11

and a long process towards resettlement and recovery, members of the community feared the worst. A distinct feature of this case is the long duration over which the damage caused by the natural hazard was distributed. Because of this feature, community members could anticipate what was to come months, if not years before the calamity would eventually be a reality. It turns out that such processes can never be anticipated accurately, however. After all, onset dynamics are not constant. Destructive processes can speed up or slow down. In the case of Aponte, both the hazard process and the resettlement planning process have taken much longer than anyone could have anticipated at their outset. Living with such anxieties (also over one's future condition) over time clearly shapes life in the present. Further, with the COVID-19 pandemic being another source of uncertainty at the time of writing it is less clear than ever when the story of the Aponte disaster can be told in its entirety.

In this book I have emphasised a number of aspects concerning the Aponte disaster to glean a few general insights for how we can approach what I refer to in this book as slow calamities. In Chapter 1 I provided relevant contextual information on the community and the disaster with the purpose of allowing readers to grasp the analytical aims of the book. In subsequent chapters, I elaborated on the Inga and their story, although this is not the main objective of the book. We saw that they had taken a vow of Wuasikamas—to protect the Earth and their territory—following a longer period of involvement in the Colombian drug trade. We also saw how these efforts led them to being awarded the UNDP-backed Equatorial Prize in 2015. I then elaborated briefly on the nature of the natural hazard phenomenon before discussing how the phenomenon is experienced. After all, how the phenomenon is experienced is more important for the topic of this book than its technical specifications. Later chapters elaborate on how the community views the slowly emerging destructive impacts of the hazard, the slow calamity as I call it, followed by reflections on how the calamity impacts fear of failed resettlement, lacking recovery and potential long-term social and economic deterioration of community. A separate chapter is also dedicated to reflecting on the analytical notion of 'life in anticipation', which I see as a potentially useful analytical lens for disaster researchers also studying other contexts and disaster types.

In this concluding chapter, I will tie together a number of loose ends from earlier in the book. Rather than being structured as one coherent narrative, the chapter is instead organised around 12 questions that arguably constitute the essence of the book. The chapter thus functions as both a summary of the entire book and its main implications, and also as a further reflection on what insights this case may offer on issues of central importance to disaster researchers and practitioners. While my objective here is to address the loose ends that remain, many of these also serve as potentially interesting themes for further research. In this research field, many conceptual (even fundamental) questions remain highly contested. One such question is what constitutes a disaster and how we may understand these phenomena (as well as the diverse ways in which those that are living them understand them). Hence, in this book I aim to tell a story, while at the same time providing fresh perspectives on a handful of central questions in the field. However, it is not my principal aim to settle any of these questions. Indeed, if the reflections below were to lead to the formulation of new questions or critiques, the field as a whole could only stand to benefit.

What characterises the experience of living in anticipation of slow calamity and/or the potential secondary disaster of resettlement?

In this book I have emphasised that ever since the gravity of the slow calamity in Aponte became apparent, life in the present was increasingly shaped by a sense of living in anticipation both of the peak intensity of the disaster and also of the resettlement process. While these two experiences are related, they are also quite different. Seeing the calamity develop slowly and worsen over time inevitably brings to mind reflections on how things will play out. Who will be affected and how? Which parts of the town will be destroyed and which will remain? How will socio-economic life be impacted? How will the livelihood of my family be affected? What about my children? Although these questions are not confined to slow calamities, the ability of affected households to slowly see their homes and those of their neighbours crumble little by little is distinct. Moreover, knowing that a significant area of Aponte would ultimately be in ruins led to an early realisation of the need for a resettlement programme. Initially, it was not clear if the relocation site would be near to the safe parts of Aponte or if it would be further away. In this book, I have also attempted to address the uncertainties and anxieties directed at the resettlement process itself, based on the way the community questioned the prospects for a full recovery in the aftermath of this disaster.

Throughout the book, I have argued, and attempted to demonstrate, that life in anticipation of slow calamity may be described as paradoxically normal. Rather than producing worldview-shattering disruptions and immediate existential risk where every second counts, slow calamities are just that——slow. During their onset, even as they worsen and produce ruins in the built environment, uproot lives and displace affected households, life still has to be lived. In a sense, slow calamities involve living despite the fact that one's surroundings are slowly disintegrating. The collective experience of readers having lived through the COVID-19 pandemic may make this affective state somewhat more relatable. Livelihoods still have to be attended to by those who can still manage such luxuries; people with caretaking responsibilities still have to carry out their work to the best of their ability; chores still need doing, to the extent that they may still be relevant; floors still need scrubbing; dinners still need cooking; and kitchens still need restocking. In this sense, slow calamity is extraordinarily similar to ordinary life.

At the same time, the lived experience of slow calamity is also a radical departure from normalcy in other ways. Although the existence of a slowly menacing threat fades in and out of conscious thinking (some may even actively attempt to distract themselves at times), such phenomena are, after all, totalising. Although slow calamities may be local or global in scope, they are always omnipresent for those living them; you can look away, you can close your eyes, but when you do look the presence of the threat remains, or may even be more advanced than when you last checked.

Throughout the entire onset of the slow calamity affecting Aponte, the Inga continued to cultivate coffee and pursue other livelihoods activities. Political and social life also endured; life throughout the onset had to continue to be lived. Still, the oftentimes paralyzing anxieties induced by the slow and steady deterioration of buildings, the land, and even central landmarks, such as the local school and church, also made it impossible to distract oneself from the reality of things for long. The lived experience of slow calamity can, in this way, be said to be characterised by a fading in and out of direct consciousness, always being there as a menacing presence, inducing dread and melancholic sentiments towards anticipated futures. The intense acuteness and immediate risk to survival that is so often associated with disasters is usually not as markedly felt. An appropriate conclusion may be that the traumas suffered are both similar and different at the same time, although the sense of 'un-ness' they trigger vary in terms of both their duration and their nature. Whilst what we may label acute disasters focus one's entire attention on the present, or the need for immediate survival, the focus of slow calamities is more future-directed, forcing exposed populations to envision and anticipate how the future is being threatened.

The sense of living in anticipation of resettlement can be thought of as a secondary disaster. Although few seem to dispute that resettlement is being used as a last resort in this case, the prospect of uprooting and uncertainties concerning the quality of the process is a source of significant anxiety and frustration for those affected. While the process is not concluded at the time of writing, it appears that the alternative site will be located near old Aponte, giving rise to some sense of continuity for those resettled. With this issue more or less settled, the protracted nature of the planning process is mainly what gives rise to the sense of life being held hostage to hiccups and delays in the process. With COVID-19 having effectively placed this issue at the bottom of the political agenda, and with budgets stripped of financing, the uncertainties surrounding the future are felt more strongly than ever. Further, the more protracted and unmanoeuvrable the relocation process becomes, the more people are at risk of displacement, either temporarily or permanently. This is directly connected to the fear concerning the survival and resilience of the community, as further displacement and further hardship strains social interactions in way that risks eroding social ties over time. Also, as people become accustomed to living elsewhere, it becomes increasingly uncertain whether or not they would return, although it can be expected that many share a strong sense of place-boundedness.

The prospect of a secondary disaster of resettlement, or failed resettlement, rather, thus gives rise to a parallel sense of future uncertainty and dread. Just as the continued development of the disaster situation always remains uncertain from the perspective of the present moment, the outcome of the resettlement process remains equally uncertain and elusive. While I can conclude on neither in this book, their implications for how we understand the temporality of disaster experience will hopefully be clear by now, in addition to the implications for affective aspects of resettlement planning. We will now turn our attention to the role that onset speed plays in more general terms.

How can we make sense of the impact that disaster onset speed may have on the lived experience of such adversities?

I have devoted considerable attention in this book to the topic of temporal aspects of the lived experience of calamity. I have argued the case that the 'onset' phase remains an understudied aspect of disaster research, perhaps mainly owing to the dismissal of onset speed as an interesting variable on the ground of two truisms in the field: (1) the observation that the rapid/slow-onset distinction is not relevant for all disasters are, in fact, slow onset, because they arise from vulnerability and not deterministically from hazards (which may have faster or slower onsets); and (2) that few hazards are associated with slow forms of destruction.

Related to the first point, think of an earthquake, a rapid-onset hazard—while it may have left a settlement in ruins in the matter of minutes, the underlying disaster risk conditions and vulnerabilities may have taken centuries to prefigure. Indeed, from the perspective of vulnerability, disasters and calamities are always longue durée processes and when analyzed in terms of precursors, not one-off events.

Related to the second point, most disasters do not allow for a careful analysis of onset. Important historical disaster occurrences, like the 2004 Indian Ocean Tsunami, the disaster that struck New Orleans in 2005, or the 2010 Haiti earthquake disaster, did not allow for research into what we may label the onset phase of the disaster due to the fact that the time frame of the disaster (unless the significantly longer, lingering aftermath is taken into account) is too short. The onset phase of most of these disasters ranged from minutes to hours to a day or two. While disasters should perhaps be seen as more than the mere destruction and immediate suffering they often cause, such as their long precursors and lasting legacies, an important point I have wanted to make with this book is that slow calamities provide us with a unique opportunity to study how communities deal with slowly worsening adversities. I argue that this is not the same as conflating hazards with disasters, as in the case of gradually occurring disasters; it is the disastrous impacts of the disaster that are manifesting slowly, not only the hazardous process underneath.

From the perspective of lived experience, slow calamities produce similar, yet different affective states. As I have argued throughout this book, disasters are normally experienced as highly disruptive, regardless of the time period over which the destruction is witnessed. While more intense disaster experiences resulting from having witnessed the world turn from the relative routine to something altogether unrecognisable in minutes or hours is distinct, generating an intense feeling of un-ness, disbelief and requiring very rapid adjustment, slow calamities are experienced as less intensely acute but still very distressing. The gradual speed at which slow calamities manifest often means that the direct threat to existence or survival is less of an experiential feature. Yet, as in, for example, the case of the Inga, communities experience the certainty of future destruction as highly distressing. The experience is not based on a theoretical possibility, but is directly observable in the form of gradually worsening cracks and damages to the built environment and the occasional collapse of buildings. It becomes an equally totalising experience as inhabitants realise that regardless of whatever one distracts oneself with momentarily, the calamity will nevertheless continue to worsen over time until little remains. The sensation of feeling powerless to stop the on-going destruction is similar, just more protracted.

Can reflections on the role of disaster onset dynamics be seen as fully compatible with the vulnerability paradigm?

The so-called vulnerability paradigm has become a unifying theoretical framework and a political movement within the field of disaster research. The perspective has largely been developed by researchers who in the 1970s were associated with the Bradford Disaster Research Unit and later popularised in key literatures in the field, such as *At Risk*, the second edition authored by Ben Wisner, Piers Blaikie, Terry Cannon and Ian Davis. Later efforts such as by the Twitter movement @ NoNatDisasters, uptake by the United Nations Office for Disaster Risk Reduction (UNDRR), the disaster studies manifesto, and several recent publications, such as JC Gaillard's 2019 paper 'Disaster Studies Inside Out' point to the maturing of the field around a relatively coherent conceptual understanding of key propositions for making sense of disaster as phenomenon. While the field was previously more divided between various disciplinary stances, such as human geography, sociology, anthropology and physical geographers, we arguably witness a gradual (albeit slow—vested interests are present also in the academy after all) convergence around this theoretical framework.

As mentioned previously, disaster researchers generally agree that there is no such thing as a natural disaster. In fact, one of the major political ambitions of the vulnerability paradigm is to speak truth to power and recognise the societal construction of disaster risk. Failure to mitigate and prepare for risks are inherently political decisions. Teasing out the agency behind the oftentimes calamitous effects of natural hazards are part of the mission behind which the research field has united. This book is very much part of this body of scholarship.

A conceptual and perhaps paradigmatic issue with which this book has grappled is how one can bring to light the need to study disaster temporalities through research centred on the onset phase of disasters. Understandably, if we break down a complex phenomenon like disaster (or calamity) into seemingly simple categories like a precursor phase, an onset phase and an aftermath phase, we clearly risk oversimplifying matters. The vulnerability paradigm, I would argue, tends to centre on the role of precursors in shaping aftermaths, and understandably pays less attention to the onset. I argued in the previous section that this probably has to do with most disasters having very short onsets. If we define the onset of a particular hurricane disaster (no, I do not mean hurricane as hazard—I refer here to a concrete disasters associated with hurricanes, such as Hurricane Katrina and the disaster it produced in New Orleans, as one possible example) as the time from which the adverse impacts of the hurricane hazard begin to manifest, until the time the havoc has been wreaked, this period of time is likely to be too short to design a study. Therefore, most disasters are studied in retrospect and not during their 'onset period'. I ask readers to forgive me for not achieving full conceptual parsimony in this section and I would argue that the difficulty of balancing the hazard/disaster distinction in these cases is interesting in its own right and worthy of further discussion among scholars in the field. As I argued in the introduction, Aponte provided a rare opportunity to study a disaster onset as it progressed slowly over time.

As we have seen, a central lesson from vulnerability theory has been the notion that no disasters are actually rapid onset. When we set out to explain disaster outcomes in terms of their precursors, it is obvious that the disaster risk creation process that led up to the calamitous aftermath took decades, if not centuries to produce. But should this render the study of 'slow-onset disasters' taboo? I would argue that distinguishing between faster and slower disaster manifestation dynamics indeed somewhat upsets the field's conceptual parsimony. Yet, and as I have tried to make clear in this book, in one way or another we need to open up a discussion about the role of impact dynamics, where the speed at which impacts manifest, is one key variable. I encourage debate on how to best achieve this conceptual 'opening up' of the field while doing as little damage as possible to the emerging conceptual coherence of the field's key theoretical frameworks. In this book I have attempted to steer clear of this debate by referring to 'slow calamity' in the book's title, but I do not consider this conceptual question to be settled by any means.

Drawing on the Aponte case I have tried to define the somewhat fuzzy borders of the hazard/disaster distinction by highlighting the difference between the onset speed of a natural hazard phenomenon and the onset speed of its resulting societal impacts. While some may object to having conceptualised disasters merely in terms of adverse impacts of hazards, two important conceptual points emerge regardless. For one, we may conceptually conceive of slowly developing hazards that produce effects that are felt nearly instantaneously (e.g. landslides characterised by very slow geological processes like slow mass movements that eventually give in and rapidly collapse, potentially producing a disaster within a very short timeframe). Conversely, we may also conceive of rapidly developing hazards that produce impacts emerging more slowly (drought could be one example, where hotter and drier conditions than normal occur when rainfall is anticipated, at first not producing very serious consequences yet with conditions worsening as time went on). Second, studying the slow intensification of disastrous conditions forces us to reflect on at which point in time the hazard transitions into a disaster. Many thresholds could be used, such as a specific death toll, the point at which emergency was declared or the moment when material damage could first be observed. However, determining when the hazard becomes a disaster is a less interesting question than considering the terminological distinction more abstractly. I would argue that engaging with this debate is by no means trivial; it has real-world consequences.

How can we make sense of the displacement dynamics of slow calamities?

The adverse impacts of natural hazards remain one of the top reasons for displacement, next to armed conflict. Actual or threatened disaster erodes communities by driving people away from their homes after livelihoods and assets have been lost, rendering life at home difficult to bear. Displacement can be either temporary or permanent, but is primarily understood in legal terms as giving rise to certain rights under national and international legal frameworks. I will not employ such an understanding of the concept in this section and I will also not be considering who is (or is not) entitled to this status. For example, national legal frameworks vary in the duration of this status. Some countries consider populations who have migrated as displaced populations for longer periods of time, whereas other countries will re-label such populations after their living situation has become more stable else-where. The notion of displaced populations is also an inherently vague category, especially in the context of elusive circumstances, as people migrate and move for such a diverse set of reasons.

Disaster-related displacement is typically understood in terms of populations seeing their homes, assets and livelihoods (basically their capacities at large) so exhausted that temporary or permanent relocation is necessary. Displacement is distinct from resettlement/relocation as it is rarely planned or organised. A very simple distinction is that displaced populations are forced to take matters into their own hands in an effort to gain a minimum of sense of security. Resettlement or relocation programmes are just that, programmes that are to some extent planned and managed (although the quality of planning and outcome varies greatly, as numerous examples from the academic literature will illustrate). I have argued in this book that the way displacement functions and is lived depends, to some extent, on the onset speed of the calamity in question.

Tangible forms of destruction and disasters with clear geographical scopes to some extent make it easier to conceptualise the difference between people directly affected and people who are affected to a lesser extent or not at all. In reality, it is more complex than this sentence suggests—disaster risk and exposure are inher-ently elusive phenomena after all—but for the sake of simplicity, it suffices to say that this is generally true. Highly elusive phenomena, slowly emerging disasters and disasters whose geographical scope is fleeting or omnipresent affect societies in a different way. I would suggest that the more elusive, gradually manifesting and ubiquitous they are, the harder attribution, meaning-making and neat categorisa-tion becomes. In this way, the displacement dynamics of slow calamities will, to some extent, blur the lines between migration and displacement and make it more difficult to distinguish direct from secondary or indirect impacts, at least from the perspective of those doing the counting and managing the fallout.

The displacement dynamics of slow calamities are mainly characterised by an absence of clearly definable tipping points after which life in a particular location becomes unbearable. The slow and incremental worsening of conditions renders us to some extent blind to the daily intensification of hardship. Yet as time goes by people will gradually exhaust their coping capacities at different rates and speeds depending on their relative position of privilege and other aspects of their life conditions that make up the total disaster risk landscape. When the displacement process is thus extended over a period of years or longer time spans, agencies may fail to take notice of every household who had to uproot their lives in order to secure their continued survival. The process is not rendered less ambiguous by the tendency for slow calami-ties to not be considered emergencies in the first place, due to their lack of perceived acuteness. After all, if people can survive until tomorrow, why respond today?

In this case study I have only been in a position to partially touch upon this highly important aspect of how slow calamities may be distinct from disasters that are felt more instantaneously. Such studies have existed for some time, and displacement is actually one of the topics most frequently studied in the context of slow-onset hazards. Droughts, famines, environmental change and the adverse impacts of climatic change have all been the subject of much displacement scholarship. Comparative approaches are still few and far between. Studying the onset phase of disasters more explicitly would allow us to discern how displacement behaves differently, depending on the duration of the onset phase. Are there big differences in how displacement works following earthquakes or floods, compared to droughts, famines or even very incremental forms of environmental change? We simply do not know, but we certainly should work to find out.

Can resettlement be considered a resilient move?

The planned and deliberate relocation of communities to an alternative site, commonly referred to as resettlement, can happen for a variety of reasons. Large-scale development projects have, over the decades, required the resettlement of scores of families. As have the unchecked disaster risk in an area, where resettlement can be carried out either pre-emptively, or reactively following a major disaster. Sometimes resettlement is also opted for if reconstruction is seen as too challenging, expensive or risky. In all of these cases, it is becoming increasingly clear that resettlement rarely improves the lives of those affected and that it should only be seen as a last resort. In those instances where resettlement remains the only viable option, outcomes depend greatly on the nature of the process leading up to resettlement, including its degree of participation, transparency, accountability and follow-up. The process involves far more than simply reconstructing the built environment elsewhere; livelihoods also have to be considered, as well as community dynamics, relative position and distance from other local communities and markets, as well as other social, economic and cultural considerations. Far too often, however, resettled communities end up in areas that are inappropriate for their livelihoods (such as fishermen being resettled to communities far inland), or to homes or areas otherwise unsuitable for their needs (homes not built according to culturally appropriate standards, a home too small for a large family). In some cases, communities may discover that the new site to which they have been relocated is exposed to unmitigated hazards.

I have in this book toyed a little bit with an admittedly very simple conceptualisation of the resilience concept——a concept commonly employed by scholars in the field. I have posed the question of whether or not resettlement can in any way be considered as compatible with how we understand resilience? In other words, is resilience mainly a place-based trait or is it more of a floating phenomenon? While I did not aim to settle this debate, I believe it to be important to problematise this much-debated concept. I mainly view this conceptual discussion as just that, a discussion that in reality has little real-world relevance for those affected or those working with them. However, given that we often expect and encourage resilience in those affected, is it necessarily so that resilience is realistic, especially if we see it as place-based?

As previously mentioned, Aponte struck me as, in many ways, a highly resilient community. Levels of inward trust is high, and they have demonstrated high levels of political innovation. Participation in decision-making is high and actively encouraged, with frequent gatherings to discuss collective matters. Women hold many posts that are central to community affairs. In other words, by most standards Aponte as a community demonstrates considerable resilience, at least if a social conceptualisation is implied. Yet the natural hazard phenomenon affecting Aponte is perpetual in nature. Disaster risk in the area will only increase with time. Slow calamity is a fact, gradual destruction is observable everywhere and quick mitigating measures are not readily available. The need for resettlement, for at least of about half of the inhabitants in the most exposed areas, is seen as a necessity. How does this bear on our understanding of Aponte as a resilient community? Or, as I phrased it above, can resettlement be considered a resilient move?

I would say that in order to conceptualise resilience in the context of resettlement we have to consider the equally contested concept of recovery. Whereas previous efforts aimed at stipulating neat sequence patterns underpinning disaster experience with reference to pre-impact-post, or mitigation, preparedness, response and recovery, the vast majority of recent scholarship recognises that these categories do not reflect social realities on the ground. Time and again we observe great heterogeneity in the way households relate to these phases, and we have also come to recognise that recovery often remains elusive even decades hence. These observations notwithstanding, I believe one of the best ways to approach the question of the place-boundedness of resilience is to consider whether recovery involves a return to the status quo or a going beyond the ex ante. Particular to the question of potential connections or inherent contradictions between the concepts of resilience and resettlement is thus whether we can make sense of the notion of 'recovering elsewhere'. Put differently, does resilience have to involve a recovery or a 'bouncing back' in the same location, or is the concept also compatible with the idea of bouncing back (or bouncing forward) in a new location?

Having posed this question, we are also forced to consider the question of whether resettlement constitutes a failure. We have already observed that relocation is generally considered only as a solution of last resort, or at least it should be. Yet we also know that not all risks can necessarily be mitigated or lived with. In some instances, exposure to disaster risks may be so unacceptably high that mitigation in situ is not a realistic or feasible option. Some adverse impacts of climatic change may, for example, render this question even more relevant. In some contexts, the exposure to the hazard will be compounded by directly attributable human actions, such as where environmental degradation and corruption or lack of enforcement have allowed for the construction of settlements on unsuitable ground, or that previously suitable ground has become unsuitable due to any of these processes. In other cases, the unbearable levels of exposure could perhaps not have been known previously due to lack of means to discover it. Either way, and regardless of how we consider the topic of agency, it seems clear that resilient communities may not always be able to withstand adversities—they may also, in some instances, be forced to recover (fully or partially, or at least try to) elsewhere. Slow calamities may simply be too difficult to live with regardless.

What constitutes community survival and what constitutes community collapse following resettlement?

Throughout the book I have reflected in various ways on the question of what constitutes community survival and what would constitute collapse following or during a resettlement process. This reflection point is directly connected to the previous one on whether or not resettlement can be considered a resilient move. I have argued that the resettlement process could at worst render the community degraded to an extent where we can question whether it survived at all, at least judging by its prior state. I have also argued that community survival or collapse is not a matter of either-or, but rather one of degree and that it may well depend on whom you ask. After all, power dynamics and spheres of influence are bound to shift during and especially following resettlement processes, particularly if any underlying tensions were brewing. In the next few paragraphs, I will reflect briefly on the question, summarise some key reflections from the book, and provide some further thoughts on the subject.

So what would constitute community collapse or at least a sense of severe failure following resettlement? Well, perhaps the indicator that most people would agree on would be a community that is far less prosperous, provides lower quality of life, lower-quality social interaction, and a lowered sense of trust and a general increase in undesirable qualities, such as lost livelihoods and mental health challenges. As we have observed, the Inga have overcome hardships before. However, one could argue that the resettlement is an altogether more contentious crisis. Previous crises that were overcome, such as organising to drive out guerilla forces, eradicating illicit crops and improving overall well-being by investing in education and governance, can be said to have functioned as experiences which strengthened the community. However, the resettlement process, having led to considerable displacement, conflict and despair, may pose a great risk to existing feelings of community.

The impact that the slow calamity and the resettlement process have on the community is likely to be affected, to a large extent, on the make-up of social dynamics after the resettlement is implemented. We have seen that aspects of urban planning are likely to play a role in interpersonal relations. If too many key social actors are displaced or otherwise lose hope and leave for elsewhere, their (unmeasurable) role in maintaining present-day community dynamics would be lost. Similarly, if the built environment is laid out in a very different way, social dynamics that depend on how neighbours and social ties are organised spatially would potentially be eroded over time. While we cannot hope for things to always be preserved and maintained without change over the years, the prospect of the deterioration of community life is a major anxiety among the Inga. Then again, perhaps their preoccupation with this matter is indicative of their strong motivation to prevent this from happening.

The perhaps biggest risk factor is the protracted nature of the resettlement process. Not only does each delay in the process contribute towards further displacement and desperation as temporary shelters have been found insufficient, but also the lack of predictability cause affected households to lose hope. With COVID-19 having shifted political priorities at higher levels, one can only imagine how the process will be impacted. Looking back and then looking forward, we can imagine that it will take time to rebuild the community in the socio-cultural sense after such a prolonged traumatic experience. Recovery will involve far more than the reconstruction of new housing near old Aponte at a safer location, it will also involve repairing the lasting effects of distrust and internal conflicts that inevitably arise under such pressing circumstances. How the community will change after the resettlement is implemented remains uncertain, but what does seem clear is that recovery will be slow and perhaps not total, at least not for all. Moreover, with each passing moment the chances for success and the time it will take for recovery in the socio-cultural sense will diminish. With low levels of inclusion, the odds of successful resettlement decrease still further. It is no wonder that the sense of anticipation that lingers over the disaster is shaped by a sense of hoping for the best while dreading the worst.

Is the management of slow calamities different?

Although the management of slow calamities shares many of the challenges associated with disaster management in general, some challenges are arguably distinct for disasters characterised by a slower onset. There are certainly many tasks that are not shaped by onset speed to a significant extent. These include, among others, the immediate task of getting most-exposed populations out of harm's way, setting up temporary shelters and clearing debris, as well as releasing funds to mitigate at least the most pressing hardships (noting that the necessary level of funding is not realistically going to be made available). After all, people who find themselves in direct danger will experience slow calamity as an acute emergency, at least in that moment in time, although the wider community will continue to experience their direct exposure as more lingering, a menace on the horizon. Some disaster management considerations are much more shaped by onset dynamics, including the onset speed of the material destruction and its social consequences.

Firstly, slow calamities will, due to their lack of perceived acuteness, receive less attention due to other competing societal issues and crises. A country, a province or a municipality will, at any one time, be facing challenges across the temporal spectrum. Lingering effects from previous scandals exercise their effects on the present and to some extent require the attention of decision-makers. Present emergencies and pressing challenges crowd up agendas. The prospects of future adversities, especially those that are perceived as relatively imminent, will also occupy the minds of those in power as well as those mandated to respond to disasters. Slow calamities will tend to be seen as issues that can wait—after all, most houses are still standing—and responses thus become more protracted and sporadic. As I noted in the introductory chapter, this often happens despite the fact that slow calamities allow for a proactive response. When impacts are known to be manifesting and people are known to be adversely affected, one would perhaps tend to think that this would mobilise precautionary action. Yet, as we observed in this book and that the experience from other slow calamities also show, the opposite often happens. Response to slow-onset adversities will thus often involve a much greater awareness-raising component, requiring considerable political entrepreneurship on the part of affected communities to attract attention to their plight and to secure at least a minimum of sustained assistance.

Second, since life, to a much lesser extent, is disrupted during the long duration that slow calamities may unfold, a socio-economic component to disaster management becomes increasingly central. As we have witnessed throughout the book, livelihoods, education, chores and daily routines somehow have to persist throughout slow calamities. Some families will naturally have their ordinary life disrupted, for example if their house is damaged to such an extent that they need sheltering or become displaced. Still, a considerable number of people will go on living through the gradual onset having to simultaneously thinking about such ordinary preoccupations as the next harvest (Aponte is a major coffee producer after all), the continuity of schooling, other local political matters, cultural activities as well as efforts to mobilise politically for increased attention to the current situation. All of these processes go on simultaneously with the gradual unfolding of the disaster, but also affect people socio-economically. For other slow-onset phenomena, such as droughts, financial instruments, insurance schemes and social safety nets are used to reduce the impacts of partially disrupted livelihoods. Being quite distinct (and indeed oftentimes seen as unrelated) from disaster management, socio-economic kinds of responses are, to some extent, better tailored to the temporal dynamics that slow calamities bring about. Of course, this does not mean that such instruments should trump other important tasks, such as a more efficient, inclusive and just resettlement process, better tending to the needs of affected families, more humane sheltering and greater community involvement.

Third, slow calamities of a perpetually worsening nature, such as the one described in this book, require studies and careful considerations about disaster risk exposure. Rigorous safety and vulnerability inquiries take time, but are necessary so as to avoid creating new disaster risk as part of the recovery effort. Responses to, and the management of, slow calamities thus allows for a more processual and well-informed set of measures. Yet, as we have seen, the outcome depends to a significant extent on sustained attention and avoiding that a slow onset serves as an excuse to delay management until more critical stages of the emergency, at which time the hardships suffered will be much greater and more lives may have been lost.

Fourth, the management of slow calamities involves a different form of disaster communication. Whereas disaster risk communication revolves around communicating to communities potential risks they may be exposed to, including the role of vulnerabilities, after-the-fact communication revolves around putting into words what has just happened. An important function of communication in the wake of disasters is thus to assist communities in conceptualising what has befallen them and to come to terms with the situation, thus functioning as the first step towards psychosocial recovery. In the case of slow calamities, however, the communication happens during the onset period of the disaster. This type of communication involves to a much greater extent a sense-making component. Uncertainties, phenomenological ambiguities and problems related to connecting the dots means that management efforts to a greater extent end up centring on raising awareness and on allowing populations to take protective actions during the slowly emerging disaster. In the absence of direct threats to life, such communication will revolve more around the minimisation of adversities than more acute forms of disaster communication, such as communication related to evacuation.

To summarise, disaster researchers have for some time tended to distinguish hazard-related response demands from disaster-related response demands. It is common sense that different kinds of hazards may produce similar kinds of consequences with implications for how they are responded to and managed. Many types of hazards demand the activation of preparedness plans, evacuation protocols, and staffing changes as well as outreach measures so as to minimise the degree to which the hazard becomes a disaster or calamity. While onset speed perhaps only to a limited extent represents a game changer among considerations in this regard, it at least greatly augments the dynamics of response demands. In the absence of intense urgency and acuteness, the pressure to act rapidly lessens; as we have seen, however, so too does political salience. Response priorities shift towards continuously securing attention, involving, among other things, efforts to declare the emerging slow calamity as an emergency (and keep it that way). After a response is secured, its nature is less focused on immediate needs and saving lives, and, in theory, more focused on reducing hardship. Maintaining the focus of authorities remains a constant struggle, and without measures, most slow calamities will ultimately become just that, calamities. A take-home lesson is that knowledge of future outcomes will not automatically lead to proactive measures in the present.

How may we conceptualise the difference between hazards and disasters in the context of slowly emerging disaster impacts?

A major conceptual obstacle to the topic discussed throughout this book stems from the way we understand disasters as processes. We have throughout this book seen how disasters are more a product of societal traits, including dynamics of vulnerability, exposure and capacities, or what we may refer to as the disaster risk creation process, than the existence of natural hazard phenomena. As noted in the introduction, speaking of slow-onset disasters is commonly seen as inconsistent with how we theorise disasters for this reason, as disaster processes are always slow and gradual; even the most rapidly occurring earthquake, thunderbolt or explosion causes harm due to unmitigated disaster risk, not merely due to their existence as natural phenomena. It makes sense, therefore, to say that hazards can have slow or rapid onsets, but that disasters are long processes per definition. Or does it?

Hazards are, in a vague sense, understood as phenomena that have the potential to cause harm. They give rise to disaster risk, but not necessarily to disasters. Some hazards occur rapidly and some occur slowly, earthquakes being an example of the former and droughts an example of the latter. Many things need to happen before disaster risks become calamities. In some ways, naming the 1990s the International Decade for Natural Disaster Reduction is fitting (if we choose to look away from it suggesting disasters are natural). By referring to disaster reduction, we can think of the end goal of disaster risk reduction as being disaster reduction—or a reduction in the severity of the disaster, with disasters not taking place at all being the ultimate reduction target (in a banal sense). Leaving disaster risk unchecked is ultimately a product of political priorities, as we know from well over half a century's worth of research into disaster causality and thinking on the subject. While conceptualising precisely the dynamics through which disaster risk creation happens is challenging, work in the field has shown that the root cause of disaster outcomes can be the result of underlying vulnerabilities that have been centuries in the making. Whether the hazard phenomenon occurs slowly or rapidly is of little relevance for an analysis of these underlying conditions, which has been among the main preoccupations of the field, at least within the vulnerability paradigm. So is any discussion of disaster onset speed inherently incompatible with how we think about disasters within this paradigm? Not necessarily.

We have seen that some hazards occur slowly and some hazards occur rapidly. The disastrous impacts of unmitigated hazards can also cause suffering and material destruction that manifests either instantaneously or slowly. This upsets the hazard/disaster distinction somewhat, except for disaster conceptualisations underpinned by what we may consider a purely vulnerability-based emphasis. By this I refer to disaster conceptualisations that explicitly or implicitly see disasters not as the adverse impacts of hazards, but as the very existence of vulnerability. In other words, there remains inconsistencies in the conceptual framework underpinning the field, and slow calamities are but one example of phenomena that reveal such inconsistencies.

We can intuitively understand that experiencing the slow and inevitable collapse of your community is not merely a hazard. We can also understand that vulnerability is a central concept also for grasping the case study described in this book. At the same time, it also opens up for reflecting on how the speed at which destruction takes place and hardships are suffered shape the phenomenology of disaster, or how disasters are experienced as phenomena. In other words, in discussing the role of disaster onset dynamics for lived experience the role of vulnerability in understanding root causes is not rendered irrelevant; it simply means applying a different angle. While I am not in a position to settle the debate on whether there exists different disaster onset speeds here in this book, I do hope that the book provides fruitful reflection points that may help us reflect further on the way in which disaster onset dynamics shape their phenomenology. Whether we refer to them as slow–onset disasters, slow calamities or gradually manifesting disaster impacts is probably of lesser importance as long as we fill our conceptualisation with substance.

Concluding reflections

In this book, I have strived to theorise how we may approach the disaster onset as a sequence distinct for slow calamities through an analytical notion I refer to as 'life in anticipation'. Some hazards may produce societal hardships that last for months, others that last for years, decades or even indefinitely. Life in anticipation of slow calamity is shaped by the effect that becoming aware of an ongoing slowly menacing disaster has on life in the present. Whether detected through direct observations or through more technical forms of measurement, hazards that are not emergencies today but that are bound to become calamities in the future have the potential to exercise a profound impact on everyday experience of disaster risk. We have also seen how such phenomena blur the boundaries between hazards and disasters (or calamities) in that it is not clear when, exactly, the hazard becomes a disaster in its consequences. If we measure calamities in terms of death tolls, perhaps they are not calamities at all? If we measure it in other ways, such as in the adversities that people suffer while living through their onsets, placing them within this category of phenomena becomes less problematic. Either way, it seems clear that ongoing, slowly developing destructive processes are distinct in their lack of acuteness and urgency, while holding a similar destructive potential. Knowledge of their future destructive potential, while perhaps not giving rise to fear for one's life and the direct risk of death, still gives rise to intense emotional responses as the future in a place becomes uncertain and the fear of not being able to recover (socially, culturally, economically) sets in. The very real possibility of a permanent reduction in an already meagre life conditions gives rise to a sense of life in the present being shaped by living in anticipation of future adversity. While the object of this anticipation can sometimes be known and experienced somewhat in the present, as is the case with slow calamities (we feel it now, but we know it will get worse later on, and things may or may not get better eventually), we have also seen that the analytical notion of life in anticipation can also be put to good use in contexts where the disaster is not in motion in the present. For example, life in a community living under a hillside where, one day, a previously unknown disaster risk potential is uncovered will also be shaped by living in anticipation of the eventual rock avalanche. Although measures may be taken to mitigate the risk, the lingering residual disaster risk will still remain, giving rise to a sense of dread for what may befall one in the future.

Dread of future calamities can thus shape lives at various geographical and temporal scales, some of them intersecting. Individually, some of us may be living in anticipation of ongoing personal adversities, such as a slowly worsening chronic illness (ongoing), or a more contingent risk, such as the existence of unmitigated disaster risk in our local environment (potential but very likely to occur one day, with potentially grave consequences). People who knowingly live on fault lines, on known floodplains or near volcanoes will also be able to relate to this analytical notion. Collectively, we experience multiple ongoing threats and narratives concerning threatening futures. Pandemics, global environmental crises, supervolcanoes, meteor strikes and the end of antibiotics are just some of examples at co-existing calamity narratives that shape life in the present, with varying time horizons and consequence potential. Anyone who has had their everyday experience impacted by such concerns, either for a time or more permanently, will be able to relate to the analytical lens I have tried to sketch out here in an admittedly crude form. The COVID-19 pandemic is another good example. When we were all told to contribute to flattening the curve in early 2020 we lived in anticipation of potential collapse—our fear being that the curve would be too steep and that we would not be able to slow the spread. As we held our breath, we saw our presents intensely shaped by fear directed at a future possibility. Regardless of how things ultimately played out in different areas of the world, the notion of life in anticipation concerns the affective state we experienced while not yet knowing the outcome. With our lives shaped by a deep sense of uncertainty and a melancholic sense of wanting to wish the looming crisis away, our lived experience was permeated by the menacing presence of the future. Whether suffered individually, as communities or collectively as a species, this affective state that I have attempted to put into words here merits further attention by scholars in the field, including fruitful critiques.

References

Adams, V., Murphy, M. and Clarke, A.E. (2009). Anticipation: Technoscience, life, affect, temporality. *Subjectivity*, *28*(1): 246–265.

Aldirch, D.P. (2012). *Building Resilience: Social Capital in Post-Disaster Recovery*. Chicago: The University of Chicago Press.

Alexander, D. (2014). Communicating earthquake risk to the public: The trial of the "L'Aquila Seven". *Natural Hazards*, *72*(2): 1159–1173.

Anderson, B. (2010). Preemption, precaution, preparedness: Anticipatory action and future geographies. *Progress in Human Geography*, *34*(6): 777–798.

Balay-As, M., Marlowe, J. and Gaillard, J.C. (2018). Deconstructing the binary between indigenous and scientific knowledge in disaster risk reduction: Approaches to high impact weather hazards. *International Journal of Disaster Risk Reduction*, *30*: 18–24.

Bankoff, G. (2004a). Time is of the essence: Disasters vulnerability and history. *International Journal of Mass Emergencies and Disasters*, *22*(3): 23–42.

Bankoff, G. (2004b). In the eye of the storm: The social construction of the forces of nature and the climatic and seismic construction of God in the Philippines. *Journal of Southeast Asian Studies*, *35*(1): 91–111.

Barrett, G. (2015). Deconstructing community. *Sociologia Ruralis*, *55*(2): 182–204.

Berents, H. and ten Have, C. (2017). Navigating violence: Fear and everyday life in Colombia and Mexico. *International Journal for Crime, Justice and Social Democracy*, *6*(1): 103–117.

Berke, P.R., Kartez, J. and Wenger, D. (1993). Recovery after disaster: Achieving sustainable development, mitigation and equity. *Disasters*, *17*(2): 93–109.

Birkmann, J., Cardona, O.D., Carreño, M.L., Barbat, A.H., Pelling, M., Schneiderbauer, S.... and Welle, T. (2013). Framing vulnerability, risk and societal responses: The MOVE framework. *Natural Hazards*, *67*(2): 193–211.

Bourdieu, P. (2000). *Pascalian Meditations*. Palo Alto: Stanford University Press.

Bruce, V. (2001). *No Apparent Danger: The True Story of Volcanic Disaster at Galeras an-d Nevado del Ruiz*. New York: HarperCollins.

Bryan, J. (2009). Where would we be without them? Knowledge, space and power in indigenous politics. *Futures*, *41*(1): 24–32.

Burnyeat, G. (2020). Peace pedagogy and interpretative frameworks of distrust: State–society relations in the Colombian peace process. *Bulletin of Latin American Research*, *39*(1): 37–52.

Chmutina, K., Sadler, N., von Meding, J. and Abukhalaf, A.H.I. (2021). Lost (and found?) in translation: Key terminology in disaster studies. *Disaster Prevention and Management*, *30*(2): 149–162.

Chmutina, K. and von Meding, J. (2019). A dilemma of language: "Natural disasters" in academic literature. *International Journal of Disaster Risk Science*, *10*(3): 283–292.

Choi, V.Y. (2015). Anticipatory states: Tsunami, war, and insecurity in Sri Lanka. *Cultural Anthropology*, *30*(2): 286–309.

Cindoy, H.C. and Chindoy, L.A.C. (2017). Wuasikamas – el modelo del pueblo inga en Aponte, Nariño (Colombia): Alli kausai o buen vivir. In: Baptiste, B., Pacheco, D., da Cunha, M.C. and Diaz, S. (eds), *Knowing Our Lands and Resources: Indigenous and Local Knowledge of Biodiversity and Ecosystem Services in the Americas*. Paris: UNESCO.

Coyle, N. (2004). The existential slap: A crisis of disclosure. *International Journal of Palliative Nursing*, *10*(11): 520.

Davis, I. and Alexander, D. (2016). *Recovery from Disaster*. Abingdon: Routledge.

DiFrancesco, D.A. and Young, N. (2011). Seeing climate change: The visual construction of global warming in Canadian national print media. *Cultural Geographies*, *18*(4): 517–536.

Displacement Solutions (2015). *Climate Displacement and Planned Relocation in Colombia: The Case of Gramalote*. Geneva: Displacement Solutions.

Fincher, R., Barnett, J., Graham, S. and Hurlimann, A. (2014). Time stories: Making sense of futures in anticipation of sea-level rise. *Geoforum*, *56*: 201–210.

Fritz, C.E. (1961). Disasters. In: Merton, R.K. and Nisbet, R. (eds), *Social Problems*. New York: Harcourt Brace & World, pp. 651–694.

Gagné, K. (2019). Waiting for the flood: Technocratic time and impending disaster in the Himalayas. *Disasters*, *43*(4): 840–866.

Garcia, A.C. (2019). *How 'the cradle of heroin' came to produce Colombia's most exclusive coffee*. Retrieved from: https://www.wuasikamas.org/news/ (February 26, 2021).

Giddens, A. (1990). *The Consequences of Modernity*. Stanford: Stanford University Press.

Giddens, A. (1991). *Modernity and Self-Identity: Self and Society in the Late Modern Age*. Stanford: Stanford University Press.

Glantz, M.H. (1999). *Creeping Environmental Problems and Sustainable Development in the Aral Sea Basin*. Cambridge: Cambridge University Press.

Granjou, C., Walker, J. and Salazar, J.F. (2017). The politics of anticipation: On knowing and governing environmental futures. *Futures*, *92*: 5–11.

Granot, H. (1996). Disaster subcultures. *Disaster Prevention and Management*, *5*(4): 36–40.

Grosz, E. (2004). *The Nick of Time: Politics, Evolution and the Untimely*. Durham: Duke University Press.

Gueri, M. and Alzate, H. (1984). The Popayan earthquake: A preliminary report on its effects on health. *Disasters*, *8*(1): 18–20.

Gundel, S. (2005). Towards a new typology of crises. *Journal of Contingencies and Crisis Management*, *13*(3): 106–115.

Hewitt, K. (1983). The idea of calamity in a technocratic age. In: Hewitt, K. (ed), *Interpretations of Calamity: From the Viewpoint of Human Ecology*. London: Allen & Unwin, pp. 3–32.

Hills, A. (1998). Seduced by recovery: The consequences of misunderstanding disaster. *Journal of Contingencies and Crisis Management*, *6*(3): 162–170.

Hulme, M., Dessai, S., Lorenzoni, I. and Nelson, D.R. (2009). Unstable climates: Exploring the statistical and social constructions of 'normal' climate. *Geoforum*, *40*(2): 197–206.

Imperiale, A.J. and Vanclay, F. (2019). Reflections on the L'Aquila trial and the social dimensions of disaster risk. *Disaster Prevention and Management*, *28*(4): 434–445.

Jimeno, M. (2001). Violence and social life in Colombia. *Critique of Anthropology*, *21*(3): 221–246.

Kelman, I. (2018). Lost for words amongst disaster risk science vocabulary?. *International Journal of Disaster Risk Science*, *9*(3): 281–291.

Kelman, I. (2019a). Axioms and actions for preventing disasters. *Progress in Disaster Science*, *2*: 1–3.

Kelman, I. (2019b). Pacific island regional preparedness for El Niño. *Environment, Development and Sustainability, 21*(1): 405–428.

Kelman, I. (2020). *Disaster by Choice: How Our Actions Turn Natural Hazards into Catastrophes.* Oxford: Oxford University Press.

Kelman, I., Mercer, J. and Gaillard, J.C. (2012). Indigenous knowledge and disaster risk reduction. *Geography, 97*(1), 12–21.

Klein, R.J., Nicholls, R.J. and Thomalla, F. (2003). Resilience to natural hazards: How useful is this concept? *Global Environmental Change Part B: Environmental Hazards, 5*(1): 35–45.

Kruke, B.I. (2015). Planning for crisis response: The case of the population contribution. In: Podofillini, L. et al. (eds), *Safety and Reliability of Complex Engineered Systems.* London: CRC Press.

Kruse, S., Abeling, T., Deeming, H., Fordham, M.... and Schneiderbauer, S. (2017). Conceptualizing community resilience to natural hazards – the emBRACE framework. *Natural Hazards and Earth System Sciences, 17*: 2321–2333.

Kuipers, E.H.C., Desportes, I. and Hordijk, M. (2019). Of locals and insiders: A 'localized' humanitarian response to the 2017 mudslide in Mocoa, Colombia? *Disaster Prevention and Management, 29*(3): 352–364.

Lewis, J. (1988). On the line: An open letter in response to 'Confronting Natural Disasters, An International Decade for Natural Hazard Reduction'. *Natural Hazards Observer, 12*(4): 4.

Li, T.M. (2007). *The Will to Improve: Governmentality, Development and the Practice of Politics.* Durham: Duke University Press.

Manyena, S.B. (2006). The concept of resilience revisited. *Disasters, 30*(4): 434–450.

Matthewman, S. (2015). *Disasters, Risks and Revelation: Making Sense of Our Times.* Basingstoke: Palgrave Macmillan.

Mazel-Cabasse, C. (2019). *Waiting for the Big One: Risk, Science, Experience, and Culture in Disaster Preparedness.* Cham: Palgrave Macmillan.

McConnell, A. (2003). Overview: Crisis management, influences, responses and evaluation. *Parliamentary Affairs, 56*(3): 363–409.

Mercer, J., Kelman, I., Taranis, L. and Suchet-Pearson, S. (2010). Framework for integrating indigenous and scientific knowledge for disaster risk reduction. *Disasters, 34*(1): 214–239.

Mika, K. and Kelman, I. (2020). Shealing: Post-disaster slow healing and later recovery. *Area, 52*(3): 646–653.

Mulligan, M., Steele, W., Rickards, L. and Fünfgeld, H. (2016). Keywords in planning: What do we mean by 'community resilience'? *International Planning Studies, 21*(4): 348–361.

Murphy, J.T. (2006). Building trust in economic space. *Progress in Human Geography, 30*(4): 427–450.

Nakagawa, Y. and Shaw, R. (2004). Social capital: A missing link to disaster recovery. *International Journal of Mass Emergencies and Disasters, 22*(1): 5–34.

Neal, D.M. (1997). Reconsidering the phases of disasters. *International Journal of Mass Emergencies and Disasters, 15*(2): 239–264.

Nixon, R. (2011). *Slow Violence and the Environmentalism of the Poor.* Cambridge, MA: Harvard University Press.

Norris, F.H., Stevens, S.P., Pfefferbaum, B., Wyche, K.F. and Pfefferbaum, R.L. (2008). Community resilience as a metaphor, theory, set of capacities, and strategy for disaster readiness. *American Journal of Community Psychology, 41*: 127–150.

Nussio, E. and Oppenheim, B. (2014). Anti-social capital in former members of non-state armed groups: A case study of Colombia. *Studies in Conflict & Terrorism, 37*(12): 999–1023.

Oliver-Smith, A. (1979). The Yungay avalanche of 1970: Anthropological perspectives on disaster and social change. *Disasters*, *3*(1): 95–101.

Oliver-Smith, A. (1991). Successes and failures in post-disaster resettlement. *Disasters*, *15*(1): 12–23.

Oliver-Smith, A. (1996). Anthropological research on hazards and disasters. *Annual Review of Anthropology*, *25*: 303–328.

Oliver-Smith, A. (2002). Theorizing disasters: Nature, power, and culture. In: Hoffman, S.M. and Oliver-Smith, A. (eds), *Catastrophe & Culture: The Anthropology of Disaster*. Santa Fe: School of American Research Press, p. 312.

Oliver-Smith, A. (2005). Communities after catastrophe. In: Hyland, S.E. and Bennett, L.A. (eds), *Community Building in the Twenty-First Century*. Santa Fe: School of American Research Press.

Oliver-Smith, A. and de Sherbinin, A. (2014). Resettlement in the twenty-first century. *Forced Migration Review*, *45*: 23–25.

Pelling, M., O'Brien, K. and Matyas, D. (2015). Adaptation and transformation. *Climatic Change*, *133*(1): 113–127.

Perry, R.W. (2018). Defining disaster: An evolving concept. In: Rodríguez, H., Donner, W. and Trainor, J.E. (eds), *Handbook of Disaster Research*, Second edition. Cham: Springer.

Perry, R.W. and Lindell, M.K. (1997). Principles for managing community relocation as a hazard mitigation measure. *Journal of Contingencies and Crisis Management*, *5*(1): 49–59

Perry, R.W. and Quarantelli, E.L. (2005). *What is a Disaster? New Answers to Old Questions*. Bloomington: Xlibris.

Radcliffe, S.A. (2017). Geography and indigeneity I: Indigeneity, coloniality and knowledge. *Progress in Human Geography*, *41*(2): 220–229.

Rosenthal, U. (1998). Future disasters, future definitions. In: Quarantelli, E.L. (ed), *What is a Disaster? Perspectives on the Question*. Abingdon: Routledge.

Sakizlioğlu, B. (2014). Inserting temporality into the analysis of displacement: Living under the threat of displacement. *Tijdschrift Voor Economische en Sociale Geografie*, *105*(2): 206–220.

Schutz, A. (1967). *The Phenomenology of the Social World*. Evanston: Northwestern University Press.

Servicio Geológico Colombiano (2016). *Informe visita técnica de emergencia resguardo indígena Inga Aponte en el municipio de El Tablón de Gómez – Departamento de Nariño*. Bogotá: Servicio Geológico Colombiano.

Shaluf, I.M. (2007). Disaster types. *Disaster Prevention and Management*, *16*(5): 704–717.

Staupe-Delgado, R. (2018). *Preparedness for Slow-Onset Disasters*. Doctoral dissertation. Stavanger, Norway: University of Stavanger.

Staupe-Delgado, R. (2019a). Can community resettlement be considered a resilient move? Insights from a slow-onset disaster in the Colombian Andes. *The Journal of Development Studies*, *56*(5): 1017–1029.

Staupe-Delgado, R. (2019b). Analysing changes in disaster terminology over the last decade. *International Journal of Disaster Risk Reduction*, *40*: 1–5.

Staupe-Delgado, R. (2019c). Progress, traditions and future directions in research on disasters involving slow-onset hazards. *Disaster Prevention and Management*, *28*(5): 623–635.

Stephan, C. and Flaherty, D. (2019). Introduction: Experiencing anticipation anthropological perspectives. *The Cambridge Journal of Anthropology*, *37*(1): 1–16.

t Hart, P. and Boin, A. (2001). Between crisis and normalcy: The long shadow of post-crisis politics. In: Rosenthal, U., Boin, A. and Comfort, L.K. (eds), *Managing Crises: Threats, Dilemmas, Opportunities*. Springfield: Charles C Thomas, pp. 200–215.

Tierney, K. and Oliver-Smith, A. (2012). Social dimensions of disaster recovery. *International Journal of Mass Emergencies and Disasters*, *30*(2): 123–146.

Titz, A., Cannon, T. and Krüger, F. (2018). Uncovering 'community': Challenging an elusive concept in development and disaster related work. *Societies*, *8*(71): 1–28.

Tulloch, J. and Lupton, D. (2003). *Risk and Everyday Life*. London: Sage Publications.

Turner, B.L., Kasperson, R.E., Matson, P.A., McCarthy, J.J., Corell, R.W., Christensen, L.... and Schiller, A. (2003). A framework for vulnerability analysis in sustainability science. *Proceedings of the National Academy of Sciences*, *100*(14): 8074–8079.

UN. (2009). *State of the World's Indigenous Peoples*. New York: United Nations.

UNDP. (2015). *Estudios de Caso de la Iniciativa Ecuatorial*. New York: UNDP.

UNDRR. (2018). Global Assessment Report 2019. Geneva: UNDRR.

van Voorst, R., Wisner, B., Hellman, J. and Nooteboom, G. (2015). Introduction to the "risk everyday". *Disaster Prevention and Management*, *24*(4): 54–58.

Waddell, E. (1977). The hazards of scientism. *Human Ecology*, *5*(1): 69–76.

Weichselgartner, J. and Kelman, I. (2015). Geographies of resilience: Challenges and opportunities of a descriptive concept. *Progress in Human Geography*, *39*(3): 249–267.

Wilkie, A. and Michael, M. (2009). Expectation and mobilisation: Enacting future users. *Science, Technology, & Human Values*, *34*(4): 502–522.

Williams, S. and Montaigne, F. (2001). *Surviving Galeras*. Boston: Houghton Mifflin.

Wisner, B. and Gaillard, J.C. (2009). An introduction to neglected disasters. *Jàmbá: Journal of Disaster Risk Studies*, *2*(3): 151–158.

Wisner, B., Gaillard, J.C. and Kelman, I. (2012). Framing disaster: Theories and stories seeking to understand hazards, vulnerability and risk. In: Wisner, B., Gaillard, J.C. and Kelman, I. (eds), *The Routledge Handbook of Hazards and Disaster Risk Reduction*. Abingdon: Routledge.

Zeiderman, A. (2016). *Endangered City: The Politics of Security and Risk in Bogotá*. Durham: Duke University Press.

Index

Pages numbers in *Italics* refer to figures; page numbers followed by 'n' refer to notes numbers.